Drinking Water Security for Engineers, Planners, and Managers

Drinking Water Security for Engineers, Planners, and Managers

Integrated Water Security Series

Mary Kay Camarillo, Ph.D., P.E.
William T. Stringfellow, Ph.D.
Ravi Jain, Ph.D., P.E.

ELSEVIER

AMSTERDAM • BOSTON • HEIDELBERG • LONDON
NEW YORK • OXFORD • PARIS • SAN DIEGO
SAN FRANCISCO • SINGAPORE • SYDNEY • TOKYO

Butterworth-Heinemann is an imprint of Elsevier

Butterworth-Heinemann is an imprint of Elsevier
The Boulevard, Langford Lane, Kidlington, Oxford OX5 1GB, UK
225 Wyman Street, Waltham, MA 02451, USA

Notice
No responsibility is assumed by the publisher for any injury and/or damage to
persons or property as a matter of products liability, negligence or otherwise,
or from any use or operation of any methods, products, instructions or ideas
contained in the material herein. Because of rapid advances in the medical
sciences, in particular, independent verification of diagnoses and drug dosages
should be made.

Library of Congress Cataloging-in-Publication Data
Application submitted

British Library Cataloguing in Publication Data
A catalogue record for this book is available from the British Library

ISBN: 978-0-12-411466-1

For information on all Butterworth-Heinemann publications
visit our web site at store.elsevier.com

Printed and bound in USA
14 15 16 17 18   10 9 8 7 6 5 4 3 2 1

Working together
to grow libraries in
developing countries

www.elsevier.com • www.bookaid.org

Contents

Preface

Securing water supplies is one of the basic functions of governance. From the beginnings of agriculture, permanent villages could only be established near reliable water sources, with the earliest known village wells dating from 6,500 BCE. The sizes of the settlements were dependent on the amount and security of the water supply and civilizations can only be located where water supplies are adequate. Bronze-age cities in the Indus Valley, Mesopotamia, and Egypt were notable for their canals and other waterworks that allowed the support of concentrated populations. The empires of Persia and Rome were characterized by aqueducts and gardens fed by sophisticated, well-engineered waterworks. Modern developed countries are characterized by a populace that assumes and expects abundant, safe water supplies.

Due to their importance, water supplies have been frequent targets of attack during periods of conflict. Early conflicts in the Euphrates basin included the diversion of whole rivers away from settlements, by the use of canals. Diocletain suppressed the rebellion of Alexandria and captured the city by cutting off the aqueducts. Conflict from ancient through to modern times frequently involved attacks on water supplies.[1] In the past decade, water security has focused on asymmetric warfare and threats from terrorist attacks, but these types of attacks are not entirely new, even in the United States, where incidents such as reservoir attacks and the Owens Valley bombings near the turn of the previous century marked times of great social and political change.

Despite the long history of water attacks, there is a new urgency to reexamine our approach to water security. Although the frequency of warfare, particularly in developed countries, may be decreasing, advances in technology, including increased global mobility and communication, have heightened the threat posed by individual and small groups. It is also increasingly recognized that water security is threatened by increasingly frequent natural disasters, including hurricanes, floods, and fires. There is increasing recognition that steps taken to increase security against deliberate contamination or destruction have dual-use value and help prepare a response to natural disasters.

Drinking water security is emerging as a crucial national concern due to the vulnerability of the water infrastructure and due to its importance to protect human health and economic wellbeing of a nation. In this book, major issues surrounding water system security are examined. This book is meant to be used as a primer by engineers, planners, and managers who work with drinking water

---

[1] See http://www.worldwater.org/conflict/ for a fascinating summary of water conflicts through the ages.

systems, as well as other water professionals. This book encompasses major aspects of water system security including: a) the current regulatory environment, b) possible threats, c) preventative measures, d) detection methods, e) emergency response, and f) system rehabilitation. To better understand the history and risk of water system contamination, recent water contamination events reported around the world were compiled and are provided in this book. Lists of available prevention, detection, and mobile treatment technologies and manufacturers are also provided as resources for creating water security plans. The emphasis is placed on the importance of planning and this book is intended to provide practical information for understanding threats and securing water systems. This book is a part of the *Integrated Water Security Technologies* book series to be published by Elsevier, which includes the complimentary volume *Low Cost Emergency Water Purification Technologies*.

Chapter 1 provides an orientation to current thinking about water security and introduces the "defense-in-depth approach" to water security that is followed throughout the book. Chapter 2 provides an in-depth look at the executive orders and legislation that have guided governmental water security initiatives since the 1990s, when the magnitude of modern terrorist threats were fully realized. Regulatory guidance, specific standards, and the EPA Response Protocol Toolbox are explained and discussed in practical terms.

The chapter on threats (Chapter 3) covers likely methods by which a water system may become compromised, with an emphasis on deliberate contamination events. Information is provided on biological, chemical, and radioactive contaminants that may be used to disrupt water supplies and the importance of understanding the physical-chemical properties of potential contaminants, in addition to toxicological properties, in evaluating potential threats and appropriate response is discussed. Prevention being preferable to response, Chapter 4 provides a breakdown of major elemental components needed for security planning. A major challenge in securing water systems is determining when a contamination event has occurred and Chapter 5 demonstrates that advances have been made in rapid detection, but that practical, on-line, or instantaneous detection of even likely contaminants is still not a reality.

How a water supplier reacts to a contaminant event can mean the difference between a nuisance and a disaster. Chapter 6 breaks down potential responses and reinforces the importance of pre-event planning, public communication, and collection of baseline information on system performance. Planning for deliberate contamination or destruction events has applicability to responding to natural disasters. Rehabilitation of systems damaged by terrorism, accident, or natural disaster is reviewed in Chapter 7. Rehabilitation can be very costly and the balance between public perception and effective mitigation strategies are considered. Systems for the emergency treatment and supply of water are reviewed and information on specific commercially available systems are provided as a resource for development of response plans.

In the final Chapter, the importance of initiative on the part of individual water supply organizations is recognized in the absence of binding regulatory requirements. Planning is the key to preventing any threat to water security and minimizing the damage from any breach of security that may occur. This book is offered as a unique, practical resource to assist in the development of adequate water security plans for large and small water suppliers. An appropriate and well-organized response will allow public health to be protected and minimize public panic and rehabilitation costs. Advances in water security have been and will be technology driven, but there are major deficiencies in current technology. Specific areas ripe for technological advancement are proposed.

The authors would like to thank Jeremy K. Domen, Stacy D. Costello, Brianna L. Juhrend, and Gordon H. Fong, graduate students at the School of Engineering & Computer Sciences, University of the Pacific, who assisted in researching and compiling data for this book. Their hard work and dedication was invaluable to the completion of this work. We acknowledge and thank Mickie Sundberg for her organizational and administrative support.

Mary Kay Camarillo,
William T. Stringfellow, Ravi K. Jain,
University of Pacific Stockton, California, USA

About the Authors

Mary Kay Camarillo, Ph.D., P.E. is an Assistant Professor of Civil Engineering and a Research Associate faculty member in the Ecological Engineering Research Program at the University of the Pacific, in Stockton, California. She received her B.S. degree in Civil Engineering from the University of Washington in Seattle, WA and her M.S. degree and Ph.D. in Civil & Environmental Engineering from the University of California at Davis. Her primary research interests are in the areas of water and wastewater treatment, agricultural anaerobic digestion, agricultural drainage management, odor control, and natural treatment systems. Prior to earning her graduate degrees, she worked in industry for seven years, working on planning, design, and construction oversight for water and wastewater treatment and conveyance facilities. She is a member of the American Water Works Association, International Water Association, and Water Environment Federation. She is a registered Professional Engineer in Washington and California.

William T. Stringfellow, Ph.D., is a Professor and Director of the Ecological Engineering Research Program at the School of Engineering & Computer Science at the University of the Pacific in Stockton, CA. He also has a joint appointment with the Geochemistry Department, Earth Sciences Division at Berkeley National Laboratory in Berkeley, CA. He received his B. S. in Environmental Health from the University of Georgia (Athens, GA) and his Master's Degree in Microbial Physiology and Aquatic Ecology from Virginia Tech (Blacksburg, VA). He received his Ph.D. in Environmental Sciences and Engineering from the University of North Carolina at Chapel Hill and worked as a Post-Doctoral Fellow in the Civil and Environmental Engineering Department at the University of California at Berkeley. Dr. Stringfellow is the first author on over 45 journal publications, has been the lead author on over 45 government reports. Dr. Stringfellow is an expert in water contamination and has made hundreds of presentations on the subjects of water quality, water treatment, and the microbiology of engineered systems.

Ravi K. Jain, Ph.D., P.E. Dean Emeritus, was Dean of the School of Engineering and Computer Science, University of the Pacific, Stockton, California from 2000 to 2013. Prior to this appointment, he has held research, faculty, and administrative positions at the University of Illinois (Urbana-Champaign), Massachusetts Institute of Technology (MIT), and the University of Cincinnati. He has served as Chair, Environmental Engineering Research Council, American Society of Civil Engineers (ASCE) and is a member of the American Academy of Environmental Engineers, fellow ASCE and fellow American Association for the Advancement of Science (AAAS). Dr. Jain was the founding Director of the Army Environmental Policy Institute, he has directed major research programs for the U.S. Army and has worked in industry and for the California State Department of Water Resources. He has been a Littauer Fellow at Harvard University and a Fellow of Churchill College, Cambridge University. He has published 17 books and more than 150 papers and technical reports.

Introduction

1. WATER SYSTEM SECURITY OVERVIEW

Protection of potable water systems is vital for protecting public health and ensuring a well-functioning society. It is important that water systems, a critical part of infrastructure, be protected from intentional damage and attack, but it is also important that thorough planning is conducted to prepare for intentional or accidental contamination. Although a large-scale attack on a water system has not occurred, concern over the safety of water systems has intensified since the terrorist attacks on September 11, 2001. An attack on a water system could result in extensive effects, including illness, loss of life, societal disruption, and economic repercussions; loss of public confidence, fear, and panic could lead to civil unrest. Similar effects can result following an accident or natural disaster. Physical attacks on water systems can occur as the result of civil disturbances, fires, vandalism, cyber attacks, breaks in main pipelines, the use of weapons of mass destruction, bomb threats, or hazardous material spills (U.S. EPA, 2008). Hazards to infrastructure, including water systems, consist of terrorist attacks, natural and human-made disasters, accidental disruptions, and emergencies (AWWA, 2009; U.S. Army CHPPM, 2005). Natural disasters can include hurricanes, tornados, floods, earthquakes, thunderstorms, winter weather, and flu pandemics (U.S. EPA, 2008).

The potential economic fallout from accidental or deliberate contamination in a water system is significant. Porco (2010) estimates that the cost for radiological contamination in a water system serving a population of 10,000 could be as high as $26 billion; for a population of 100,000, the estimated economic impact could be $100 billion. These cost estimates took into account immediate effects, such as loss of life, cessation of industrial activities, emergency medical care, rapid response and security personnel, and remediation, as well as secondary effects, including effects to sanitation systems, disruption of firefighting capabilities, long-term economic disruption, long-term medical care, decreases in property values, and public fear and distrust. In the National Infrastructure Protection Plan (NIPP), four types of consequences are defined: public health and safety, economic, psychological, and governance effects (AWWA, 2009).

The security of a water system can be compromised in a variety of ways (DeNileon, 2001). The physical components of the water system can be directly attacked, such that water system components cease to function or function incorrectly. A cyber attack would target the computer systems used to operate the water infrastructure, causing similar malfunctions. A third type of attack could come in the form of deliberate contamination. Potential repercussions of intentional contamination are similar to potential repercussions from accidental contamination. Contamination arising from accidents could be the result of a natural disaster, accidental mistakes in operating the water system, or an accident such as a train derailment (e.g., chemicals from a derailed train are discharged into a water body that serves as a drinking water source). Preparing for and detecting accidental and deliberate contamination are similar and parallel activities. It is important that water utilities adopt an "all hazards" approach to emergency planning, so that they prepare for all types of adverse situations.

Understanding the potential contaminants is critical in preparing for water system failures and attacks. Many biological, chemical, or radiological agents are available and could be used to harm or frighten a significant population via the water system. Many contamination agents can be acquired in sufficient quantities to cause casualties to a large population. To some extent, drinking water systems are protected against biological contaminants as a result of maintained concentrations of disinfectant residual within the distribution system. Biological attacks are believed to be most dangerous in the aerosol form, and most research has focused on the possibility of an airborne attack or event. However, biological contaminants can be dispersed in drinking water systems, and many potential biological agents are immune to disinfection with chlorine. One difficulty in protecting potable water supplies is that many potential chemical contaminants are not removed in conventional treatment systems. In addition, deliberate contamination could occur in the distribution system downstream of treatment facilities. Understanding the nature of potential contaminants is crucial for determining how best to detect the contaminants, respond to an attack, and rehabilitate the portions of impaired water systems exposed to contaminants.

Water utilities can take many steps to ensure water security. Linville and Thompson (2006) describe the following as features of an effective water security program:

- Management must be committed to security.
- Vulnerabilities must be assessed and monitored.
- Investments to improve security should be given priority.

- Access to facilities should be controlled.
- Access to security-sensitive information should be controlled.
- Security considerations should be incorporated into emergency response planning.
- Communication strategies should be implemented.
- Security status should be maintained.

AWWA (2009) states that there should be a security culture throughout water utility organizations, where employees have defined roles and expectations. Documents should be kept current and continuously revised with resources dedicated to security activities and improvements. A lot of effort has been dedicated in the last ten years toward development and testing of contaminant warning systems; these systems are very useful in making water system more secure (U.S. EPA, 2010). Additional steps that water utilities can take include the incorporation of security features into the design and construction of new facilities, development of protocols for responding to threats, and formation of partnerships with other agencies, public health officials, and law enforcement.

Preparing for accidental and deliberate contamination requires knowledge in several areas. First, potential threats must be identified. Also, possible prevention measures must be determined and implemented. Detection technologies and strategies for identifying contamination are useful tools for reducing the impact of contamination. Finally, information regarding potential responses to contamination and rehabilitation methods for affected systems must be collected to be adequately prepared. Water providers must determine their system vulnerabilities and plan for emergency situations. *The purpose of this book is to review the issues surrounding water security and provide information useful to engineers, planners, and managers to address and manage these challenges in order to enhance water security.*

2. HISTORICAL PERSPECTIVE

The possibility of a terrorist attack on U.S. water systems is well established (NRC, 2002). Throughout the world and throughout history, attacks on drinking water have been carried out as a military and terrorist tactic (Beering, 2002; Christopher et al., 1997; Gleick, 2006). In particular, the use of pathogenic microorganisms and biotoxins as weapons has a long history throughout the world and is a real modern threat.

Throughout history intentional attacks have been made on water supplies, resulting in casualties and damage to water systems and, more important, causing distress to the affected population. In a comprehensive review of water intrusions, Gleick (2006) suggests that it is relatively difficult to incur widespread damage to water systems but that the societal effect, including the induction of panic and distrust in the government, is potentially very significant. There are even biblical references to contamination of drinking water wells as a war tactic (Winston and Leventhal, 2008). In more recent times, intentional contamination has occurred, although the attacks are not always carried out on drinking water supplies. Highly publicized attacks have occurred through food and aerosols. Some of the significant threats to water security since 1941 are listed in Table 1.1.

In addition to the events summarized in Table 1.1, there have been many smaller-scale attacks on water systems, in many cases motivated by vandalism. Based on an American Water Works Association (AWWA) study, it appears that small-scale threats to water systems are common, but many of these threats do not result in actual contamination of the water system (Nuzzo, 2006). During the preparation of this book, newspaper articles were collected from an English language newspaper database from December 2005 to January 2011 to characterize threats to water systems (Appendix B). As a result of the search,

TABLE 1.1 Summary of Significant Threats to U.S. Water Systems Since 1941

Date (reference)	Threat
January 2001 (DeNileon, 2001)	The FBI received a threat from a North African terrorist group warning of their intention to disrupt the water supply in 28 U.S. cities. Although the FBI determined that it was not a viable threat but a forgery, the Association of Metropolitan Water Agencies (AMWA) had already been notified of the threat and distributed a memorandum to all water providers serving 100,000 customers or more.
July 2002 (Cameron, 2002)	The FBI confiscated documents of U.S. water systems along with information regarding water poisoning from an Al-Qaeda suspect in Denver, Colorado.
February 2003 (CDC, 2003)	The Centers for Disease Control (CDC) and U.S. Environmental Protection Agency (U.S. EPA) issued a "Water Advisory in Response to High Threat Level." While the advisory was not in response to a specific threat, the purpose of the advisory was to inform the public about the potential risk of a terrorist attack.
May 2003 (*Washington Times*, 2003)	Representatives of Al-Qaeda informed an Arabic language newsmagazine that they had plans to poison water systems in U.S. and other Western cities.
2004 (Kalil and Berns, 2004)	The FBI and DHS issued a warning regarding a plot to poison a water supply during chlorination.

nearly 80 events were identified from locations throughout the world. The actual number of events is probably much higher, since only events that were reported in the news media were located. Of the water system threats reported, many of the threats consisted of planned attacks that were not carried out. Of the attacks that did occur, most involved use of readily available materials, such as fertilizers and rat poison.

In addition to intentional attacks resulting from terrorism or vandalism acts, water systems can be compromised as the result of natural disasters and accidents. Information obtained from all water system threats can be used to learn about system vulnerability and develop better approaches to strengthen water security. Experiences from natural disasters can also be helpful in for-mulating agency-specific emergency response procedures. For example, in the aftermath of Hurricane Katrina in 2005, panic among the affected population arose from shortages of water and other critical supplies (Brodie et al., 2006). Lessons can be learned from this disaster and other events to assist in pre-paring for future disasters. Water providers can develop emergency response guidelines that are appropriate for different types of water system attacks and failures.

3. DEFENSE-IN-DEPTH APPROACH

Water security is an extensive issue, requiring a systematic approach for plan-ning, implementing improvements, and measuring success. To enhance water security, the U.S. EPA has adopted a "layered defense" or defense-in-depth approach to securing water systems. The defense-in-depth approach is intended to mitigate risk by employing a protective, best-available management strategy by providing a multitude of devices to prevent attacks, including contaminant warning systems to become informed once an attack has occurred and deploy-ment responses once an attack has been confirmed (Ashley and Jackson, 1999). These protective measures include physical barriers and hardware, security devices, communication protocols, operational response procedures, and per-sonnel screening (DeNileon, 2001). The defense-in-depth approach is advo-cated in published guidelines for development of comprehensive water security programs (AWWA, 2009).

One of the most basic ways that water providers can improve water security is to make adjustments in the way they operate and maintain their water systems. In general, a well-maintained system and well-managed agency is more resistant to attack. For example, water system personnel that are well trained and alert are more likely to notice and react to suspicious situations. Development of a "security culture," where employees are aware, can greatly improve water secu-rity, but this culture must be endorsed by management (Linville and Thompson, 2006). Magnuson et al. (2005) presents a list of actions that water providers can do to prepare for an intentional attack on a water system. Some of the actions involve little or no technology investment, such as providing additional security training for employees. Working with law enforcement agencies is an important step to improving water security (DeNileon, 2001). Establishment of communication systems is important for preparing to respond to an event; in particular, it is important that employees understand these communication mechanisms (Linville and Thompson, 2006). In addition, water providers need to balance the sharing of information with the public, as required by the Freedom of Information Act (FOIA), with the need to ensure water security and protect information regarding critical infrastructure of such water systems (DeNileon, 2001; Haimes, 2002). For example, one of the authors of this book was easily able to obtain detailed records of water supply infrastructure for a local municipality using an online public bidding website.

The defense-in-depth approach to water security can be applied to water security by understanding vulnerabilities and threats, planning aggressively, training for emergencies, installing physical security barriers and equipment, being prepared to respond to a threat, and understanding how a recovery effort would take place. Personnel issues, training, cyber security, and other similar issues, while important to water security, are outside the scope of this book. The focus here is on: planning to evaluate vulnerabilities and potential threats (Chapters 2 and 3), physical prevention measures (Chapter 4), contaminant monitoring and threat evaluation systems (Chapter 5), timely response protocols for contamination events (Chapter 6), and efficient rehabilitation of affected water systems (Chapter 7).

4. DUAL-USE BENEFITS

There are no regulatory requirements for specific water security improvements, and these improvements can be costly (Copeland, 2010). Therefore, it is important that water utilities adopt water security improvements that have benefits in addition to providing water security.

An example of a dual-use benefit is the use of online contaminant monitoring systems for improving distribution system water quality. Monitoring the water quality in distribution systems can be done to establish baseline conditions, so that significant changes in water quality can be detected (Smeti et al., 2009). Online data collection can be used to detect cross-connections, identify valve operation problems, make better operational decisions, and minimize disinfection by-product concentrations by optimizing chlorine dosage (U.S. EPA, 2007). Online contaminant monitoring can also provide dual-use benefits by accomplishing routine sampling, monitoring system pressure, and detecting system leaks (Hart and Murray, 2010). Hagar et al. (2013) found that event detection software offers dual-use benefits to improve operations and treatment decisions, leading to improved disinfectant residual, biofilm control, and drinking water aesthetics (taste, color, and odor). Using online water quality monitors is also more efficient than having to send operators out to remote sites to collect data and samples (Kroll and King, 2010; Schlegel, 2004). The benefits of online sensors in distribution systems has become more evident as issues with aging infrastructure become more significant.

An additional dual-use benefit of water security is that water utilities improve their system modeling capabilities. Developing contaminant event detection software requires dynamic (not static) water distribution modeling software that provides a dual-use improvement, because dynamic programs can also be used for planning purposes (Roberson and Morley, 2005). Also, information technology (IT) and supervisory control and data acquisition (SCADA) systems must be updated to accomplish water security goals. There are also benefits of merging distribution system water quality data onto existing SCADA systems (Serjeantsonet al., 2011).

Development of comprehensive emergency planning and training is useful in preparing for all types of emergencies, including flu pandemics and earthquakes. Consequence management plans should cover all types of emergency situations. Planning can result in improved coordination with response organizations and partnership with other agencies as well as improved laboratory capabilities and awareness of external laboratory resources (U.S. EPA, 2007). Installation of physical barriers at facilities can reduce losses occurring from vandalism and theft. Improved consumer complaint systems can result in better customer service. Employee training can also be beneficial in achieving a better functioning and more efficient organization. The added benefits that result from providing a more secure water system must be emphasized because the benefits of water security alone are not easily measured or appreciated. For example, the success of a secure water system is evident by the absence of attacks or intrusions in that system.

5. TECHNOLOGY SOLUTIONS

Use of the defense-in-depth approach to water security suggests that attention be paid to planning for intentional attacks, employing physical prevention devices, contaminant warning systems, response systems and protocols, and rehabilitation strategies, so that water systems can be repaired efficiently and returned to service or quickly abandoned. All of these layered defenses are based on adaptation and use of technology. Technology development is an important component of the overall goal of making drinking water safer (NRC, 2002). Technology can be used to accomplish the defense-in-depth approach to water security; the infrastructure needed to physically protect facilities and detect contamination relies on technology. The U.S. EPA and other agencies have been actively working on research and development of technologies to assist water utilities in achieving better, more secure water systems (U.S. EPA, 2004a, 2004b). Developing new technology and applying it is an important task for the water industry. In addition, funding and support to develop this technology is needed (U.S. GAO, 2003, 2004). The additional infrastructure needed to detect and potentially respond to intentional attacks on water systems is being adapted slowly because of the lack of available funds and the absence of regulations requiring water providers to install such infrastructure (Nuzzo, 2006).

CONCLUSIONS

Water system security is vital for protecting public health and ensuring a well-functioning society, as the potential economic and psychological fallout from accidental or deliberate contamination of a water system is significant. The history of intentional attacks suggests that water system security could be compromised in a variety of ways, and a terrorist attack on U.S. water systems is a possibility.

Water security is an extensive issue, requiring a systematic approach for planning, implementing improvements, and measuring success. Although water utilities can take many steps to ensure water security, the most basic approach is to make adjustments in the way that they operate and maintain their water systems, along with implementing a security culture, where employees have defined roles and expectations. The defense-in-depth approach to water security can be applied by understanding vulnerabilities and potential contaminants, planning aggressively, training for emergencies, installing physical security barriers and contaminant warning equipment, being prepared to respond to a threat, and having rehabilitation strategies for efficient repair and the ability to return to normal operating conditions.

Due to the lack of regulatory requirements for specific water security improvements, water utilities are encouraged to implement systems, such as online contaminant monitoring systems and system models, with dual-use benefits for both water security and other aspects of water system management. The development of comprehensive emergency planning and training is useful in preparing for not only contamination events, but other natural disasters and emergencies that may affect water system operation. Clearly, enhancing water security is not just about responding to deliberate attacks; planning for emergencies is a basic requirement of infrastructure management. Although a large amount of information on water security is available, the need remains to organize and summarize the tools and resources available to make them more accessible to water professionals.

REFERENCES

American Water Works Association (AWWA), 2009. Security practices for operation and management. ANSI/AWWA G430-09. AWWA, Denver, CO.

Ashley, B.K., Jackson, G.L., 1999. Information assurance through defense in depth. IA Newsletter 3, 3–6.

Beering, P.S., 2002. Threats on tap: Understanding the terrorist threat to water. J. Water Resour. Plann. Manag., ASCE 128 (3), 163–167.

Brodie, M., Weltzien, E., Altman, D., Blendon, R.J., Benson, J.M., 2006. Experiences of Hurricane Katrina evacuees in Houston shelters: Implications for future planning. Am. J. Publ. Health 96 (8), 1402–1408.

Cameron, C., 2002. Feds arrest Al Qaeda suspects with plans to poison water supplies. Fox News July 30, 2002.

Centers for Disease Control and Prevention (CDC), 2003. CDC health advisory: CDC and EPA water advisory in response to high threat level. Department of Health and Human Services CDC, Atlanta, GA.

Christopher, G.W., Cieslak, T.J., Pavlin, J.A., Eitzen, E.M., 1997. Biological warfare, a historical perspective. JAMA- J. Am. Med. Assoc. 278 (5), 412–417.

Copeland, C., 2010. Terrorism and security issues facing the water infrastructure sector. Congressional Research Service. Washington, DC.

DeNileon, G.P., 2001. The who, what, why, and how of counterterrorism issues. J. Am. Water Works Assoc. 93 (5), 78–85.

Gleick, P.H., 2006. Water and terrorism. Water Policy 8, 481–03.

Hagar, J., Murray, R., Haxton, T., Hall, J., McKenna, S., 2013. Using the CANARY event detection software to enhance security and improve water quality. Environmental and Water Resour. Institute (EWRI) of ASCE, Cincinnati, OH 1–14.

Haimes, Y.Y., 2002. Strategic responses to risks of terrorism to water resources. J. Water Resou. Plann. Manag.- ASCE 128 (6), 383–389.

Hart, W.E., Murray, R., 2010. Review of sensor placement strategies for contamination warning systems in drinking water distribution systems. J. Water Resour. Plann. Manag.- ASCE 136 (6), 611–619.

Kalil, J.M., Berns, D., 2004. Drinking supply: Terrorists had eyes on water, security bulletin reveals treatment facility plot details. Las Vegas Review-Journal. August 12, 2004.

Kroll, D., King, K., 2010. Methods for evaluating water distribution network early warning systems. J. Am. Water Works Assoc. 102 (1), 79–89.

Linville, T.J., Thompson, K.A., 2006. Protecting the security of our nation's water systems: Challenges and successes. J. Am. Water Works Assoc. 98 (3), 234–241.

Magnuson, M.L., Allgeier, S.C., Koch, B., De Leon, R., Hunsinger, R., 2005. Responding to water contamination threats. Environ. Sci. Technol. 39 (7), 153A.

National Research Council (NRC), 2002. Making the nation safer: The role of science and technology in countering terrorism. NRC, Washington, DC.

Nuzzo, J.B., 2006. The biological threat to US water supplies: Toward a national water security policy. Biosecurity Bioterrorism-Biodefense Strategy Pract. Sci. 4 (2), 147–159.

Porco, J.W., 2010. Municipal water distribution system security study: Recommendations for science and technology investments. J. Am. Water. Works. Assoc. 102 (4), 30–32.

Roberson, J.A., Morley, K.M., 2005. Contamination warning systems for water: An approach for providing actionable information to decision-makers. American Water Works Association (AWWA), Denver, CO.

Schlegel, J.A., 2004. Automated distribution system monitoring supports water quality, streamlines system management, and fortifies security. J. Am. Water Works Assoc. 96 (1), 44–46.

Serjeantson, B., McKenny, S., van Buskirk, R., 2011. Leverage operations data and improve utility performance. AWWA Opflow 37 (2), 10–15.

Smeti, E.M., Thanasoulias, N.C., Lytras, E.S., Tzoumerkas, P.C., Golfinopoulos, S.K., 2009. Treated water quality assurance and description of distribution networks by multivariate chemometrics. Water Res. 43 (18), 4676–4684.

U.S. Army Center for Health Promotion and Preventive Medicine, 2005. Emergency response planning for military water systems, USACHPPM TG 297. Water Supply Management Program, Aberdeen Proving Ground, MD.

U.S. Environmental Protection Agency (U.S. EPA), 2004a. EPA's role in water security research: The water security research and technical support action plan, EPA/600/R-04/037. EPA, Washington, DC.

U.S. Environmental Protection Agency (U.S. EPA), 2004b. Water security research and technical action plan, EPA/600/R-04/063. EPA, Washington, DC.

U.S. Environmental Protection Agency (U.S. EPA), 2007. Water security initiative: Interim guidance on planning for contamination warning system deployment, EPA 817-R-07-002. EPA, Washington, DC.

U.S. Environmental Protection Agency (U.S. EPA), 2008. Decontamination and recovery planning—Water and wastewater utility case study, EPA 817-F-08-004. EPA, Washington, DC.

U.S. Environmental Protection Agency (U.S. EPA), 2010. Water quality event detection systems for drinking water contamination warning systems, EPA/600/R-10/036. EPA, Washington, DC.

U.S. General Accounting Office (U.S. GAO), 2003. Drinking water: Experts' views on how future federal funding can best be spent to improve security, report to the Committee on Environment and Public Works, U.S. Senate. GAO-04-29. Washington, DC.

U.S. General Accounting Office (U.S. GAO), 2004. Drinking water: Experts' views on how federal funding can best be spent to improve security, testimony before the Subcommittee on Environment and Hazardous Materials, Committee on Energy and Commerce, House of Representatives. GAO-04-1098T, Washington, DC.

Washington Times, 2003. Al Qaeda warns of threat to water supply. Washington [DC] Times. May 28, 2003.

Winston, G., Leventhal, A., 2008. Unintentional drinking-water contamination events of unknown origin: Surrogate for terrorism preparedness. J. Water Health 6, 11–19.

Chapter 2

U.S. Regulatory Environment and Planning for Water Security

1. LEGISLATION

A series of Executive Orders, Presidential Decision Directives (PDDs), Homeland Security Presidential Directives (HSPDs), Presidential Policy Directives (PPDs), and other legislation has strengthened the U.S. stance on security issues. Much of this effort has been in response to the terrorist attacks on September 11, 2001. However, even before this egregious event, security was a significant concern for U.S. policy makers. This concern prompted a series of presidential directives and legislation that directly affected the operation and design of water systems.

1.1. Executive Order 13010, July 15, 1996, Critical Infrastructure Protection

In this executive order, the water supply is identified as one of the eight national infrastructures vital to security, along with telecommunications, electrical power systems, gas and oil storage and transportation, banking and finance, transportation, emergency services, and continuity of government. Executive Order 13010 introduced the President's Commission on Critical Infrastructure Protection (PCCIP), which was formed in recognition of the need for government to collaborate with the private sector and to develop a strategy for protecting critical infrastructure. The general objectives of the PCCIP, as established by the executive order, are as follows:

- To identify and collaborate with entities of the public and private sector that contribute to the operation of critical infrastructure.
- To assess the nature and extent of vulnerabilities and threats to critical infrastructure.
- To determine what legislative issues may be of concern by efforts to protect critical infrastructures and determine how these issues should be handled.
- To recommend a comprehensive national policy and implementation plan for protecting critical infrastructures from threats and assuring their continued operation.
- To propose statutory or regulatory changes required to effect the recommendations.
- To produce reports and recommendations as they become available, that is, not limit itself to producing one final report.

1.2. PDD63, Critical Infrastructure Protection, May 22, 1998

This presidential directive was developed in recognition that advances in information technology had created new vulnerabilities in critical infrastructure with regard to equipment failure, human error, weather, and other natural causes. It is established in this directive that these vulnerabilities require a flexible and evolving approach to protect critical infrastructure and enhance national security. The importance of public–private partnerships in reducing potential vulnerabilities was emphasized.

The National Infrastructure Protection Center (NIPC) was established and was located within the Federal Bureau of Investigation (FBI) to address the concerns covered in PDD63. The Critical Infrastructure Assurance Office (CIAO) was also created. The NIPC serves as headquarters for national critical infrastructure threat assessment, warning distribution, and vulnerability review. The NIPC is linked electronically with federal agencies as well as public and private entities involved with critical infrastructure operations.

As part of PDD63, the Information Sharing and Analysis Center (ISAC) was developed, which functions as a pathway for representatives from the private sector and federal government officials to share and distribute information regarding vulnerabilities, threats, and intrusions to critical infrastructure. The Water Information Sharing and Analysis Center (WaterISAC), was established, whereby water utilities can become members and receive information about water threats. The WaterISAC provides rapid notification to its members that are located in the United States, Canada, and Australia. The database maintained by WaterISAC is centralized, kept up to date and secure, ensuring timely distribution of information. Notifications are sent via secure electronic transmission and are intended to warn water utilities of potential or perceived threats, so that they can respond accordingly (by increasing surveillance, sampling, etc.). The Water ISAC is managed by the Association of Metropolitan Water Agencies (AMWA).

Within the directive, the U.S. EPA is identified as the lead agency for water security, and the Center for Disease Control (CDC) is identified as the lead agency for public health. Following the passage of this directive, the U.S. EPA formed a partnership with the American Metropolitan Water Association, which was critical in developing better lines of communication with managers of public water systems. The partnership between the U.S EPA and AMWA led to the development of the Critical Infrastructure Protection Advisory Group (CIPAG), a group of experts providing input on water system vulnerabilities and threat assessment.

1.3. Homeland Security Act, 2002

The Homeland Security Act of 2002 established the Department of Homeland Security (DHS), a federal agency charged with overseeing all security-related activities including acts of terrorism, natural and human-made crises, emergency planning, and other illegal activities that may threaten domestic security. Among its myriad of responsibilities, the DHS is required to carry out comprehensive assessment of the vulnerabilities of the critical infrastructure, including water systems, in the United States. As part of this act, the functions of the NIPC and CIAO were transferred to the DHS.

1.4. Public Health Security and Bioterrorism Preparedness and Response Act (PL 107-188) (Bioterrorism Act), 2002

The Bioterrorism Act was passed to improve the ability of the United States to prevent, prepare for, and respond to bioterrorism and other public health emergencies. It requires water utilities serving more than 3300 people to complete vulnerability assessments (VAs) for their water systems to determine how susceptible these systems are to terrorist attack and other malicious acts intended to disrupt the ability of the system to provide a safe and reliable supply of drinking water. The VAs are required to be submitted to the U.S. EPA administrator and are not made available to the public. The VAs are required to include, at a minimum, a review of all water system components as well as the operation and maintenance procedures. As a result of the Bioterrorism Act, public water utilities serving populations of over 3300 people are also required to prepare and update emergency response plans (ERPs). The ERP is required to include plans, procedures, and identification of equipment that could be utilized in the event of an intentional attack on the water system and actions, procedures, and identification of equipment that could be used to mitigate the impact of an intentional act on the water system. The ERPs are intended to incorporate information and results found during the preparation of the VAs. Although not required, smaller systems are also encouraged to complete VA and ERP documents. The Bioterrorism Act does not specifically require security improvements or that any of the vulnerabilities identified in the VA be addressed. No specific guidelines or standards were developed for ensuring better security of water systems.

Passage of the Bioterrorism Act led to the development of the National Infrastructure Protection Plan (NIPP), a strategy developed by DHS to coordinate security efforts across multiple sectors. The NIPP is based on a risk management framework. Critical infrastructure components are defined and classified into various sectors, with roles and responsibilities assigned for the protection of each sector. Each sector was assigned a lead agency and these agencies are responsible for developing sector-specific plans (SSPs). The U.S. EPA was identified as the lead sector-specific agency for water infrastructure protection. Implementation of the NIPP and the sector-specific organizational structure led to the development of a water sector-specific plan prepared by both DHS and the U.S. EPA and containing guidance to assist water utilities in being better prepared for terrorist attack and other hazards (U.S. EPA, 2008a). In the NIPP, four types of consequences for attacks on water systems are identified: public health and safety, economic, psychological, and governance effects. Defining the four types of consequences is useful in assigning risk associated with various threats.

In addition to establishing SSPs, the NIPP led to formation of sector coordinating councils (SCC) that are comprised of representatives from the specific sectors. The purpose of the SCCs is to facilitate communication among those involved in the sector and support information sharing. The SCCs address policy issues regarding critical infrastructure protection planning, make recommendations to improve water security, and advise government agencies on sector-specific security issues, such as planning, research and development, and policy. The SCC for the water sector is the Water Sector Coordinating Council (WSCC). The WSCC reviews policy, water security strategies, and recommends actions to reduce water system vulnerabilities. The WSCC comprises representatives from the American Water Works Association (AWWA), Water Research

Foundation, Association of Metropolitan Water Agencies (AMWA), National Association of Clean Water Agencies (NACWA), National Association of Water Companies (NAWC), National Rural Water Association (NRWA), Water Environment Federation (WEF), and Water Environment Research Foundation (WERF). The WSCC advocates a tiered approach to conducting risk assessments that includes taking the following potential effects into consideration: loss of life, economic effect, psychological effect/continuity of government, and critical customers (U.S. EPA, 2008a).

1.5. HSPD-5, Management of Domestic Incidents, 2003

Enactment of HSPD-5 tasked the DHS with developing and administering the National Incident Management System (NIMS), which includes a water sector NIMS. The purpose of NIMS is to provide a nationwide approach for federal, state, and local governments to prepare for, respond to, and recover from domestic incidents. The motivation for the NIMS structure is for responders of different jurisdictions and disciplines to coordinate planning efforts and work together during a crisis situation. The basis for NIMS was the National Interagency Incident Management System (NIIMS), a pre-existing management system. The directive HSPD-5 also led to development of a national response plan (NRP), which includes protocols for operating under various threats and consistent approaches to reporting incidents. As part of NIMS, an incident command system (ICS) was developed that could be adapted to all of the critical infrastructure sectors.

1.6. HSPD-7, Critical Infrastructure Identification, Prioritization, and Protection, 2003

This presidential directive resulted in a comprehensive national policy for federal entities to identify, prioritize, and protect critical infrastructure and key resources. This directive reinforced the U.S. EPA's role as the lead federal agency for drinking water and water system security. This directive effectively replaced PDD63, which had been issued in 1998. In HSPD-7 and other documents, VAs are referred to as *risk assessments* (AWWA, 2010).

1.7. HSPD-9, Defense of United States Agriculture and Food, 2004

This presidential directive resulted in a national policy intended to defend the agriculture and food systems against intentional attacks, significant disasters, and other emergencies. As part of the directive, the U.S. EPA is charged with developing surveillance and monitoring programs that could provide early detection of water contamination events. Additionally, HSPD-9 required the U.S. EPA to develop and improve intelligence operations and analysis capabilities of information concerning threats and delivery systems relevant to the water sector and to collaborate with other federal agencies to develop mitigation, response planning, and recovery strategies for an attack, disease outbreak, or other disaster affecting agriculture and food infrastructure.

Passage of this directive led to development of the U.S. EPA WaterSentinel program (later known as the *Water Security Initiative*) to improve water security surveillance and monitoring by developing and evaluating contaminant warning systems (CWS) to detect contamination events in distribution systems quickly (U.S. EPA, 2007c, 2008d–f). The initiative was carried out in three phases: the conceptual design of a CWS, demonstration of a pilot CWS, and guidelines for implementing new CWSs. The pilot CWS was implemented and tested in Cincinnati, Ohio, to demonstrate the functionality of CWS and refine the approach for implementation in a full-scale water system. The pilot CWS consisted of the following monitoring and surveillance components: online water quality monitoring, sampling and analysis, enhanced security monitoring, consumer complaint surveillance, and public health surveillance. Subsequent pilot CWSs were funded for New York City and San Francisco. Another function of the Water Security Initiative was to develop guidance on consequence management plans (U.S. EPA, 2008f).

Directive HSPD-9 led to formation of the National Drinking Water Advisory Council (NDWAC) (NDWAC, 2005). The NDWAC is a group of stakeholders formed by the U.S. EPA to provide input on a variety of regulatory, guidance, and policy issues related to drinking water, including water security. The NDWAC provides insight on numerous drinking water issues by forming working groups to address specific issues. The Water Security Working Group (WSWG) was formed to facilitate "the development of voluntary best security practices" that the group recommended to the NDWAC (NDWAC, 2005). The goals of the WSWG are to:

- Establish the features of effective security programs.
- Identify water utilities to encourage effective programs.
- Suggest ways to measure progress at the utility level and nationally.

To support the contaminant monitoring and potential responses to threats, the U.S. EPA was directed to develop a nationwide laboratory network that integrates existing federal and state laboratory resources. These efforts to distribute information regarding analytical capabilities led to development of the Water Laboratory Alliance (WLA), a nationwide network of laboratories that are qualified to perform many analyses for chemical warfare agents, bioterrorism agents, and radiological constituents (Antley and Mapp, 2010). The WLA is a water infrastructure component of the Environmental Response Laboratory Network (ERLN). A comprehensive ERLN/WLA response plan provides guidance regarding the laboratories in the network (U.S. EPA, 2010c, 2012b).

1.8. HSPD-10, Biodefense for the Twenty-First Century, 2004

This directive addressed biological warfare agents, including limiting and controlling the development of new agents, developing worldwide surveillance systems, producing countermeasures such as vaccines and medications, and other activities. This directive resulted in the development and deployment of biodetection technologies and decontamination methodologies for water infrastructure.

1.9. PPD 8, National Preparedness, 2011

This policy directive replaced HSPD 8, which had addressed national planning. This directive required agencies such as the U.S. EPA to develop plans to demonstrate preparedness for "acts of terrorism, cyber-attacks, pandemics, and catastrophic natural disasters." This directive also required federal agencies to coordinate their planning and preparation activities.

2. REGULATORY REQUIREMENTS AND GUIDANCE

The U.S. EPA has responded strongly in its role as the lead federal agency in the United States for water security (U.S. EPA, 2004b). To meet the objectives of the presidential directives and legislation discussed previously, the U.S. EPA received directed funding and established a number of programs, alliances, and guidance documents to assist public water suppliers in making their systems more secure. The role of the U.S. EPA in water security involves research and technical support, information exchange, tools and protocols, financial assistance, incorporating security into the water industry, emergency response, and training (U.S. EPA, 2004b). Much of the research and development work completed has been accomplished by the U.S. EPA Office of Water's Water Security Division in collaboration with the National Homeland Security Research Center.

The federal government can play an important role by providing guidance to water utilities and the data needed to determine appropriate procedures and responses (Nuzzo, 2006). The U.S. EPA developed the following goals regarding water security: (1) sustain protection of public health and the environment; (2) recognize and reduce risks; (3) maintain a resilient infrastructure; and (4) increase communication, outreach, and public confidence (U.S. EPA, 2008a). The U.S. EPA goals are reflected in the efforts made to provide guidance and support to water providers. The U.S. EPA advocates an "all hazards" approach to planning and emergency preparedness.

Security of water systems is an extensive problem, as water systems span large geographic areas, have a multitude of significant physical components, and many entry points. Further, an attack on a water system could come in the form of a direct attack (i.e., physical obstruction), cyber attack on the computerized control system, or a deliberate contamination intrusion. A physical attack on water infrastructure could include destruction of a physical component of the system or disruption to the power supply, communications system, chemical system, or supervisory control and data acquisition (SCADA) system (DeNileon, 2001). Water systems are extensive and many portions cannot be adequately controlled (e.g., plumbing in individual residences). Significant vulnerabilities include water facilities in parks and open spaces with public access, public tours of treatment plants and other key facilities, and uncovered water reservoirs containing treated water (Linville and Thompson, 2006).

2.1. Regulation of Drinking Water Systems

The water supply in the United States is controlled by multiple public and private agencies operating at the local, state, and federal levels. To some degree, the decentralization of water systems is an asset in protecting these systems, as it is difficult to attack these fragmented and diverse systems. However, the decentralized nature of water systems means that water security is the responsibility of multiple organizations. Although the U.S. EPA provides guidance, there are currently few regulatory requirements for water providers to address water security in the operation of a water system. Also, there are no specific standards for ensuring security in water systems (Copeland, 2010). To some extent, existing drinking water regulations and monitoring requirements provide some level of water security. Water providers routinely monitor a host of constituents in their water supplies and report the results to relevant state agencies and their consumers. However, the monitoring methods used are insufficient in the number of constituents monitored and the analyses performed to adequately detect a deliberate contamination event in a timely manner.

2.1.1. Drinking Water Standards

The passage of the Safe Drinking Water Act (SDWA) in 1974 allowed the U.S. EPA to regulate and enforce drinking water standards in the United States. Prior to the passage of the SDWA, states were responsible for regulating drinking water systems and usually followed guidelines provided by the U.S. Public Health Services (Roberson, 2011). The SDWA required the U.S. EPA to enact National Primary Drinking Water Regulations, which set mandatory maximum contaminant levels (MCLs) for over 80 contaminants in drinking water that can cause negative health effects in the general public (U.S. EPA, 2009). Regulated contaminants include microorganisms, disinfectants, disinfection byproducts, organic and inorganic chemicals, and radionuclides. The U.S. EPA also provides maximum contaminant level goals, which are below MCLs and provide a factor of safety for the public health, but are unenforceable (U.S. EPA, 2009). States have the power to enforce their own set of regulations on drinking water providers, as long as they meet or exceed the standards set by the U.S. EPA. In addition to MCLs, the U.S. EPA sets sampling schedules and testing methods, along with standard procedures for treating water with contaminants that exceed the MCLs.

In addition to the National Primary Drinking Water Regulations, the U.S. EPA established National Secondary Drinking Water Regulations. Secondary standards are not mandatory and act as guidelines for drinking water systems to manage contaminants that pose no threat to the public health but negatively affect the aesthetic, cosmetic, and technical effects of drinking water (U.S. EPA, 2012a). Public drinking water systems can test for secondary contaminants on a voluntary basis to ensure levels do not exceed the secondary maximum contaminant levels, as high levels may cause the public to stop using the water regardless of it being safe to drink (U.S. EPA, 2012a). Contaminants that cause aesthetic effects, such as changes in odor, taste, and color, as well as those that cause cosmetic effects, such as skin or tooth discoloration, are more likely to reduce the public usage of drinking water. Contaminants that cause technical effects, such as corrosion, scaling, and sedimentation, can have significant economic effects on a public water system.

The SDWA requires the U.S. EPA to develop a contaminant candidate list (CCL) every five years, and make decisions on whether to update drinking water regulations to include the contaminants on the list. The CCL is used to help guide scientific studies and data acquisition on the potential health effects and treatment technologies in order to determine if regulation of a contaminant is needed. CCL 3, published in 2009, contains 104 chemicals and 12 microbial contaminants that are currently unregulated but may be regulated in the future (Appendix C).

2.2. Regulatory Guidance

Multiple guidance tools and documents for improving water security are available from the U.S. EPA and other sources; however, none of these are binding. Instead, the U.S. EPA's Water Security Initiative encourages the nationwide adoption of additional water security measures, as opposed to implementing additional requirements for the drinking water industry. While some regulations concern water security, most are focused on vulnerability assessment and response planning rather than minimum security standards, detection systems, and treatment technologies.

2.2.1. Recommendations from the Water Sector Working Group

Following passage of HSPD-9, the Water Sector Working Group (WSWG) was convened to discuss water security issues, and this group then presented their findings (NDWAC, 2005). The WSWG represented a diverse group of representatives from the Water Sector; it was comprised of representatives from public and private utilities, large and small water and wastewater utilities, public health agencies, environmental regulators, and public interest groups. The final report published by the NDWAC stressed that water security in the United States is based on voluntary programs and not on regulations. There is some disagreement within the water sector community regarding the adoption of water security regulations. Without regulations, it is difficult for many water utilities to justify the expenditures associated with water security infrastructure, especially given the expense of such systems and the lack of tangible results and benefits. However, the proponents state that a regulatory approach would be difficult to implement, since each water utility is different and utility-specific approaches to water security make the most sense.

The WSWG offered 18 main findings based on its work to identify water security issues. One of the most important findings of the WSWG was the identification of characteristics of an effective water security program. They found that knowledgeable employees and institutional preparedness were very important and that these result from strong leadership that values security. Water utilities with good security programs invest resources in planning, establish clear procedures, conduct periodic reviews, keep documentation up to date, incorporate security features into new and existing facilities, form partnerships, and assess program effectiveness. The final report suggests minimum standards for ensuring water security, which serves as a checklist for water utilities, as well as a list of security measures that is useful for water utilities.

In conducting vulnerability assessments, water utilities should consider system failures that include water pressure loss, loss of supply, chemical releases, public health effects from contamination events, loss of wastewater treatment, and use of the wastewater and storm water collection systems to stage attacks (NDWAC, 2005). In addition, water utilities should consider threats that include physical disruption of service, intention or accidental contamination, cyber attacks, and the use of conveyance infrastructure to carry out other attacks (NDWAC, 2005). Contaminants used in an attack can include biological, chemical, and radiological agents.

The WSWG had several suggestions for state and federal agencies to assist water utilities (NDWAC, 2005). They suggested that U.S. EPA, DHS, state agencies, and water/wastewater utility organizations promote water security as an important issue and provide information to water utilities. The same groups could provide peer technical assistance, review of programs, and assistance locating resources. Information intended for the public and public officials (e.g., city councils, rate setting organizations) would be helpful for water utilities that are trying to educate their customers on the importance of water security improvements and the necessity of investing in such improvements. Many professional organizations (e.g., American Society of Civil Engineers) give awards for successful projects; one suggestion made by the WSWG was to offer awards in the area of water security. Finally, the WSWG recommended that grant and loan programs be established to help defray the costs of water security improvement projects.

2.2.2. Response Protocol Toolbox

To prepare for contamination events and develop the most appropriate responses, water providers are advised to use the U.S. EPA Response Protocol Toolbox (RPTB), which is a rigorous planning tool to aid in the creation of agency-specific response guidelines that serve as a "field guide" for responding to contamination events (Magnuson et al., 2005; U.S. EPA, 2004c). The response guidelines in the RPTB should be incorporated in the emergency response plan of the water provider and referenced during training exercises. Additional response preparation guides are available to assist water providers (WA DOH, 2005). Although the RPTB was produced by a federal agency, it is not a regulatory document and contains general information that should not be used during an event. Instead, water utilities need to use the RPTB to develop their own system-specific documents to plan and formulate system-specific responses. For example, water utilities should document contact information and appropriate procedures in their ERPs. The RPTB provides information on available technology for detecting water quality anomalies, presenting attacks, and responding to specific threats, although it does not make specific technology recommendations. Additional guidance is provided from the U.S. EPA in the form of a handbook, which is a condensed and more accessible version of the RPTB (U.S. EPA, 2006). The RPTB contains six modules (Table 2.1); the contents of these modules are discussed further in Chapter 5.

TABLE 2.1 U.S. EPA Response Protocol Toolbox Modules (U.S. EPA, 2003c)

Component	Title	
Module 1	Water Utility Planning Guide	Provides a discussion of the nature of the contamination threat to the water supply and describes the planning activities a utility can conduct to manage contamination incidents
Module 2	Contaminant Threat Management Guide	Provides a framework for evaluating threats and making decisions regarding the appropriate actions to take in response to the threat
Module 3	Site Characterization and Sampling Guide	Describes the process of collecting information from the site of a suspected contamination event, including site investigation, field safety screening, field testing, and sample collection
Module 4	Analytical Guide	Provides a framework to developing an approach for the analysis of samples collected from the site of a suspected contamination event
Module 5	Public Health Response Guide	Examines the role of the water utility during a public health response action as well as the interactions between the utility and other parties involved (e.g., primary water agency, public health community, public officials)
Module 6	Remediation and Recovery Guide	Describes the planning and implementation of the remediation and recovery activities required to restore the water system prior to resuming normal operation after a contamination event

2.2.3. Guidance Tools and Documents

The U.S. EPA produced several additional guidance documents and resources to aid water providers in ensuring their water system is safe and secure, including those described in Table 2.2.

TABLE 2.2 Regulatory Guidance Tools Established by the U.S. EPA to Enhance Security of Water Systems

Guidance Tool	Description
Climate Resilience Evaluation and Awareness Tool (CREAT)	A software tool to help water utilities comprehend and adapt to potential threats and risks associated with climate change effects (e.g., flooding, drought, water quality) to water utility assets. This tool is also helpful in evaluating water security related to potential intentional contamination events.
Community Based Water Resiliency (CBWR) Tool	An electronic tool that provides communities with the ability to assess their water providers' resiliency against natural disasters, aging infrastructure, and intentional contamination while raising awareness of interdependencies between water utilities and community services (e.g., emergency response efforts) that are critical to water system resiliency.
Effective risk and crisis communication during water security emergencies (U.S. EPA, 2007a)	A report summarizing research results from water security risk communication message mapping workshops. These results assist water utilities in predicting questions that the media and public may ask following a contamination event and provide information to help water utilities prepare clear and concise answers to these questions.
Environmental Laboratory Compendium	An online database of environmental laboratories containing each laboratory's specific capabilities to analyze chemical, biological, and radiological contaminants, developed to address the limited analytical capabilities that most water providers have in terms of monitoring contaminants.
Developing consequence management plans for drinking water utilities (U.S. EPA, 2008f)	The consequence management plan defines a process for establishing the credibility of a suspected incident, the response actions that may be taken to minimize public health and economic consequences, and a strategy to restore the system to normal operations if a contamination event has been identified.
Developing an operations strategy for contamination warning systems (U.S. EPA, 2008e)	This document provides guidance to utilities for the development of recommended standard operating procedures for the monitoring and surveillance components of a contamination warning system. The guidance focuses on development of an operational strategy that integrates the components to provide a timely indication of a possible contamination event.

TABLE 2.2 Regulatory Guidance Tools Established by the U.S. EPA to Enhance Security of Water Systems—cont'd

Guidance Tool	Description
National Environmental Methods Index for Chemical, Biological and Radiological Methods (NEMI-CBR)	A searchable database of analytical methods for contaminants of concern and appropriate responses.
Sampling guidance for unknown contaminants in drinking water (U.S. EPA, 2008c)	Procedures for sample collection, preservation, and transport to support multiple analytical approaches for the detection and identification of potential water contaminants.
Tabletop Exercise Tool for Water Systems (TTX Tool) (U.S. EPA, 2011a)	A CD-ROM-based tool allowing the water sector to conduct and evaluate tabletop exercises that allow the utilities to practice, test, and improve emergency response plans. Contains scenarios to address a comprehensive set of threats (e.g., natural disasters, human-made incidents, climate change) and includes a situation manual, discussion questions, and a PowerPoint presentation to aid in the tabletop exercise execution.
Vulnerability Self-Assessment Tool (VSAT) (U.S. EPA, 2010b)	A software tool to assist water utilities of all sizes assess security risks and threats and update emergency response plans. The steps involved in the risk assessment include (1) analysis setup and utility information, (2) asset identification, (3) countermeasure evaluation, (4) threat identification, (5) baseline assessment, (6) proposed new countermeasures, (7) cost/risk evaluation, and (8) summary report.
Water Contaminant Information Tool (WCIT) (U.S. EPA, 2007b)	A password-protected online database that provides information regarding contaminants of concern for water security to help utilities make appropriate response decisions if their water system is threatened by accidental or intentional contamination. The database can be accessed by water utilities, state agencies, federal officials, public health agencies, and water associations. The contaminants of concern include chemical, biological, and radiological constituents that would be detrimental if introduced into drinking water.
Water Health and Economic Analysis Tool (WHEAT)	A software tool that assists drinking water utilities to quantify health and economic effects of a potential water system attack. The current version of WHEAT can analyze two scenarios: introduction of hazardous gas and loss of operating assets in a drinking water system. Future models will analyze more contamination scenarios for both water and wastewater.

Continued

TABLE 2.2 Regulatory Guidance Tools Established by the U.S. EPA to Enhance Security of Water Systems—cont'd

Guidance Tool	Description
Water Laboratory Alliance Training Center (U.S. EPA, 2012b)	The WLA provides the water sector with a nationwide network of laboratories to support water contamination events. The training center provides the water sector with information regarding response procedures, analytical methods, sample handling recommendations, and data reporting through computer-based training modules and live training at conferences.
Water utility response, recovery, and remediation actions for man-made and/or technological emergencies (U.S. EPA, 2002a)	This document provides guidance for response, recovery, and remediation activities water utilities can undertake when threatened by a human-made or technological emergency. This guide contains various scenarios of contamination types and provides a list of recommended persons and entities to notify in a case of such threats as well as specific actions that can be commenced.

In addition to producing guidance documents to assist water providers in preparing for contamination events, the U.S. EPA is actively involved in research and development efforts. These efforts are conducted at a variety of locations, including the Homeland Security Research Center. Work is also conducted by the Distribution System Research Consortium and as part of the Water Security Standards Program, Threat Ensemble Vulnerability Assessment (TEVA) Program, Environmental Technology Verification (ETV) Program, and other programs. For example, one contribution made by the U.S. EPA is the development of a water distribution model, PipelineNet, which is used to predict the transport and fate of contaminants in distribution systems. This model can also be used to determine optimal placement of sensors (Chapter 4). The U.S. EPA has expended considerable effort in developing and testing contaminant warning systems, which was mandated in HSPD-9.

3. PLANNING EFFORTS AND DOCUMENTATION

Water utilities need to develop rigorous risk assessment and emergency preparedness plans so that they are prepared for all types of hazards. These plans should include water security considerations so that the water utilities are prepared for all types of malevolent acts, including water contamination threats and events. Water utilities need to develop vulnerability assessments, response plans, notification procedures, internal chains of commands, and schedules for implementing improvements to reduce vulnerability. Water utilities also need to conduct training for their employees and hold exercises where employees practice responding to emergency events. Currently, the primary bases for water security-specific efforts made by water utilities are in the development of their VA and ERP documents. The U.S. EPA provides guidance in preparing these documents (U.S. EPA, 2003a, b, 2004a), as do many state agencies (Crisologo, 2008; WA DOH, 2003).

Water security planning may reveal that water utilities need to make significant investments to improve security and reduce risk. In evaluating vulnerabilities, preparing for potential threats, and training personnel to respond to potential attacks on water systems, it is important that water providers allocate funds to address system vulnerabilities. Water utilities are expected to fund improvement projects using rates and connection fees. No direct funding is available from the federal government for water security improvements, although communities with water systems that have undergone an attack may be eligible for federal disaster relief funding. The federal government has responded to these shortcomings and has allocated funds to develop and field test new technologies to improve water security and allocated funds for states to operate training centers for small water providers (U.S. GAO, 2003, 2004). Copeland (2010) reports that the federal government spent $923 million to support assessment and risk reduction activities and to protect federal facilities; however, federal funding had not been allotted for improvement projects. States have some resources to assist water providers in assuring that their systems are resilient to attack (Crisologo, 2008).

Providing secure water systems is a huge undertaking for U.S. water providers. Some agencies involved in water supply are large organizations employing an extensive staff of highly trained employees that oversee extensive facilities and operations. Other water providers are small organizations that may be remotely located and provide potable water to small communities. Developing a comprehensive framework for emergency planning is important and should include cooperation with regulatory agencies, public health officials, other water utilities, and other groups involved in emergency planning (Jalba et al., 2010). To ensure water security for the diverse water systems located throughout the United States, it is important that the individual water providers use the U.S. EPA guidance documents and develop their own system-specific approaches and plans. Each water system has unique vulnerabilities (e.g., remote supply, densely populated areas, aging infrastructure, location in a hurricane zone) that must be identified and addressed. In all cases, water providers should be knowledgeable of the potential threats (e.g., biological warfare agents) that could be used in an attack.

The Water/Wastewater Agency Response Network (WARN) was formed by the AWWA to assist water utilities in understanding regulatory requirements and resources available. Membership in WARN is voluntary and the intent is for utilities to share information and form partnerships, potentially signing mutual aid agreements. WARN establishes best practices for specific regions and often holds meetings and training sessions on a variety of emergency response and planning topics.

3.1. Vulnerability Assessments

Water utilities serving more than 3300 individuals are required by the Bioterrorism Act to prepare VA reports to evaluate susceptibility to potential threats and provide a plan for upgrades, modifications, changes in operational procedures, and policy changes to mitigate risks (AWWA, 2001; U.S. EPA, 2002b). Although not required, these documents are also recommended for small water utilities serving fewer than 3300 individuals. Although the U.S. EPA collects VAs from more than 8000 water utilities throughout the country, it is prevented from sharing any of the collected information as a result of confidentiality, in accordance with the Bioterrorism Act (Nuzzo, 2006).

3.1.1. Components of a Vulnerability Assessment

An important part of the VA is for water providers to understand and document their water system components as well as the potential for adverse events to occur and the anticipated repercussions (Table 2.3). Documentation should include inventories of facilities, locations, and sizes of water system features. Documentation should also take the form of developing a computer model of the entire water supply and distribution system that integrates demand and operations management features. Personnel should be trained in understanding system components. Water providers should also develop a good understanding of their customers. For example, some customers may be targets of attack or more vulnerable to repercussions suffered during a water contamination event (e.g., hospitals). Water providers should also have a good understanding of locations within the water system that are vulnerable to the introduction of contaminants (e.g., remote sites). They should understand the availability of chemicals within their service areas (e.g., industrial facilities, storage facilities for agricultural chemicals).

3.1.2. Sources of Threat Information

Vulnerability assessments should be continually updated to reflect potential threats (biological, chemical, and radiological). Water utilities should periodically review information provided by the U.S. EPA such as the data contained in the databases WCIT and WHEAT. The Water Contaminant Information Tool (WCIT) is a secure database containing information about potential threats that can assists water utilities with VAs and ERPs. Access to the WCIT is password-protected, and water utilities are eligible for access. The WaterISAC can also assist water utilities with preparation of VA and ERP documents in addition to fulfilling their role in issuing water threat announcements.

3.1.3. Water Distribution System Models

Establishment of a water distribution system model is an important step in preparing for a contamination event. Water providers should have detailed plans of their systems and should have the system simulated using software

TABLE 2.3 Vulnerability Assessment Components (U.S. EPA, 2002b)

Component	Description
1. Characterization of water system	Identify utility-specific information, including: • Customers served • System location (service area map) • Inventory of assets and facilities (e.g., pipelines, constructed conveyances, impoundments, pretreatment facilities, treatment facilities, storage structures, electronic infrastructure, computer and automation systems, chemical systems) • Identify interdependence of assets
2. Description of potential consequences	Identify consequences, including: • Potential effects and disruptions • Extent of effects • Public health effects • Economic effects • Effects on social peace and public confidence
3. Determination of critical assets that could be subject to attack	Identify possible scenarios, including: • Physical damage or destruction • Contamination of water • Chemical releases • Interruption of electricity • Interruption of communication systems Attacks on all critical assets should be considered, including distribution systems, storage, electrical systems, controls systems, chemicals, and treatment systems.
4. Assessment of likelihood of attacks	Identify possible modes of attack and likelihood, which is a difficult task. Consider using: • Design Basis Threat (DBT) evaluation process • U.S. EPA's "Baseline Threat Information for Vulnerability Assessments of Community Water Systems" publication • Information from the WaterISAC
5. Evaluation of existing countermeasures	Identify methods to prevent and mitigate threats, including: • Intrusion detection systems • Water quality monitoring • Operational alarms • Guard orders • Employee training and exercises • Physical barriers (locks, fencing, restrictions on vehicle access, etc.) • Cyber protection • Security procedures and policies (personnel procedures, badge system, chemical deliveries, package deliveries)
6. Analysis of risk and development of a plan to reduce risk	Accumulate all information from the review into an overall assessment of risk and recommended actions that would reduce risks by reducing vulnerabilities. Methods to reduce risks fall into three categories: sound business practices, system upgrades, and security upgrades.

that can be used for modeling system conditions as well as for determining locations affected during a contamination event to guide response, rehabilitation, and recovery efforts. The following software packages are used by water providers for modeling their distribution systems: EPANET, MWH Soft H2ONET, KYpipe PIPE2000, Advantica Stoner SynerGEE Water, WaterCAD, PipelineNet, and DH1 Software Mike Net. Having good working knowledge of the distribution system is a fine way to prepare for responding to a contamination event effectively. Jeong et al. (2009) used hydrodynamic and water quality models to simulate radiological contamination of the drinking water source for Seoul, South Korea. Skolicki et al. (2008) used sensor network software to examine different combinations of attack scenarios with preventative measures to determine which activities were most protective.

3.1.4. Water Distribution System Maintenance

One way that water providers can prepare to respond to contamination events is to maintain their distribution systems. For example, installation of isolation valves throughout the system is a good idea and can assist in response efforts. For example, the Ten State Standards provides guidance on valve spacing (www.10statestandards.com). Regular flushing and pigging of pipelines and cleaning of reservoirs can minimize biofouling and scaling that could negatively affect the impact of a contamination event by retaining contaminants in the system. Additionally, selection of appropriate pipe materials may affect the degree of fouling and the difficulty in cleaning system components.

3.1.5. Risk Assessment Frameworks

To address the complexity and importance of water security, it is advisable for water providers to manage and mitigate risks, such that decision making is optimized. Risk management offers an approach to understanding risks and the ability of facilities to undergo attack, allowing for better decision making by water utility managers and regulators (Morley, 2010). There are many methodologies for applying risk management strategies, and protocols have been developed specifically for water systems. In conducting risk assessments, it is necessary to consider costs as well as benefits, since efforts to reduce vulnerability are often limited by available technology and economics.

The idea of using risk management to address water security has gained momentum. Haimes (2002) describes a hierarchical holographic modeling methodology, which includes modeling of not only water systems but also other influencing factors, such as the terrorist organizations themselves and the geopolitical climate. Matalas (2005) states that it is difficult to view terrorist acts as probabilistic, since there is an element of surprise and they are not regularly occurring events. Sekheta et al. (2006) recommends a hazard analysis and critical control points (HACCP) approach to threats in food and water supplies. Khan et al. (2001) also propose HACCP as a potential tool for water security risk management, although the authors identified surveillance as the key to identifying contamination events. In applying risk assessment frameworks, the possible scenarios must be determined as well as the likelihood of these events and the potential consequences (Haas, 2002).

Although difficult, it is not impossible to determine the risk of hazards occurring. Several tools are available for assisting water utilities in preparing their VA documents; most of these tools are based on risk assessment methodology:

- The Vulnerability Self-Assessment Tool (VSAT) developed by the National Association of Clean Water Agencies
- The Security and Environmental Management System (SEMS) software developed by the National Rural Water Association (NRWA)
- The Risk Assessment Methodology—Water (RAM-W) developed by a partnership between the Water Research Foundation, Sandia National Laboratories, and the U.S. EPA
- The Risk Analysis and Management for Critical Asset Protection (RAMCAP®) process (ANSI/ASME-ITI/AWWA J-100) developed by the American Society of Mechanical Engineers (ASME) and the American Water Works Association (AWWA)

The VSAT (Vulnerability Self-Assessment Tool) is risk assessment software that organizes the vulnerability assessment components recommended by the U.S. EPA (U.S. EPA, 2010b). The VA documents should be updated periodically to reflect system changes and upgrades, and the VSAT tool enables utilities to continuously improve to ensure more secure water systems. The SEMS software is specifically intended for systems serving between 3,300 and 10,000 customers.

The AWWA standard, RAMCAP®, contains a methodology for analyzing and managing risks associated with both natural hazards and attacks to potable water systems (Table 2.4). The RAMCAP® tool provides guidance on calculating the probability of attacks and hazards as well as guidance on calculating asset and utility resilience, which was missing from previous vulnerability and water utility risk assessment methods (AWWA, 2010; Morley, 2010). Risk (R) is calculated using the following equation:

$$R = CVT \tag{2.1}$$

where C is the consequences in terms of number of fatalities and serious injuries, V is the vulnerability, or the likelihood that a threat will result in the consequences, and T is the likelihood of the threat occurring as a probability or frequency over a given time period (AWWA, 2010; ASME-ITI, 2006). Utility resilience is defined as the ability of a water utility to withstand attacks and effects from natural disasters. Resilience of a water system asset (R_a) is calculated using the following equation:

$$R_a = DSVT \tag{2.2}$$

TABLE 2.4 Seven Steps for Conducting a Vulnerability Assessment Using the RAMCAP® Methodology

Component	Description
1. Asset characterization	• Identify and prioritize critical water system assets based on the worst reasonable consequences
2. Threat characterization	• Determine and categorize threats as human-made hazards (intentional or accidental), natural hazards, and dependency hazards • Determine the range or magnitude of each threat from the smallest that would cause serious harm to the largest reasonable case • Apply threat characterization to each critical asset
3. Consequence analysis	• Determine the worst reasonable consequences, as measured by number of fatalities and injuries, financial loss to facility owners, economic effect to community, political effect, psychological effect, environmental effect, public confidence, etc.

4. Vulnerability analysis	• Analyze critical asset ability to withstand threats
	• Determine likelihood of consequences if a hazard (e.g., human-made, natural, or dependency) occurs.
	  • Review system details
	  • Analyze vulnerability of each asset using event-tree analysis, path analysis, vulnerability logic diagrams, computer simulation methods, etc.
	  • Document method(s) for performing the vulnerability analyses
	  • Determine likelihood of attack

5. Threat analysis	• Calculate likelihood and frequency of threats
6. Risk/resilience analysis	• Calculate water system risk and resilience for each threat-asset couple. Risk is quantitatively defined as the product of the consequences (C), vulnerability (V), and threat (T).
7. Risk/resilience management	• Determine acceptable levels of risk and resilience • Identify improvements to reduce risk • Evaluate cost-benefit ratio for each proposed improvement

where D is the duration of service denial, in days, and S is the severity of service denied, in gallons per day (Biringer et al., 2013). The owner's economic resilience (R_{oe}) can be estimated using the following equation:

$$R_{oe} = R_a P \qquad (2.3)$$

where P is the predistribution unit price of service (Biringer et al., 2013). The community economic resilience (R_{ce}) can be estimated using the following equation:

$$R_{ce} = EVT \tag{2.4}$$

where E is the economic loss to the community (Biringer et al., 2013). In addition, in the RAMCAP® methodology, vulnerability from both natural hazards and terrorist attacks are considered, which is valuable, since preparing for natural disasters and terrorist attacks can usually be done simultaneously.

Tidwell et al. (2005) offer a systematic approach for calculating overall asset security for individual water system elements to withstand specific attacks using a Markov latent effects (MLE) modeling approach. Two calculations can be used to calculate overall asset security, a soft-aggregation weighted sum (Eq. 2.5) and a simple linear weighted sum (Eq. 2.6):

$$WS_k = \frac{1}{1 + e^{-5.5\left(\sum_{i=1}^{n} w_i x_i + \sum_{j=1}^{m} v_j y_j - 0.5\right)}} \tag{2.5}$$

$$WS_j = \sum_{i=1}^{n} w_i x_i + \sum_{j=1}^{m} v_j y_j \tag{2.6}$$

where WS_k and WS_j are the weighted sum for the kth and jth decision elements respectively, w_i and v_j are corresponding weights, n and m are the number of attribute values, while m is equal to any latent effects that contribute to decision elements, x_i is the attribute values input to individual decision elements, and y_i is the latent effect inputs due to prior elements (Tidwell et al., 2005). Using the MLE approach, Tidwell et al. (2005) assign values to system attributes that define the degree of security provided (e.g., effectiveness of physical barriers). Next, values are calculated for decision elements (e.g., overall physical security) using (Eqs. 2.5 and 2.6). Asset security is then calculated considering the weighted sum of all of the values for the decision elements, such that the overall asset security represents all the elements and attributes that are influential (Tidwell et al., 2005).

3.2. Emergency Response Plans

Emergency response plans contain information that is needed and procedures that should be followed in the event of an emergency (U.S. EPA, 2003a, 2004a). The ERPs are essential planning documents, because in the event of an emergency, the response needs to proceed very quickly. Water utilities are required to prepare ERPs and keep the information contained within these ERPs current. These ERP documents have long been used to address water system preparedness for dealing with natural disasters and accidents, and there are many similarities between responses to natural disasters and deliberate events, such as terrorist attacks (AWWA, 2001; U.S. Army CHPPM, 2005). Water systems serving more than 3300 people are required to address information from their VAs in their ERPs, specifically intentional contamination events. Jesperson (2002) provides a summary of commonsense advice for water providers, which stresses the importance of planning and the role of participation by all involved individuals, including the need to establish communication among employees so that all employees and emergency response personnel understand the response protocols.

3.2.1. Components of an Emergency Response Plan

According to the Bioterrorism Act, ERPs should contain plans and procedures that would be carried out in the event of an attack. The ERPs should also contain identification of resources, personnel, assisting agencies (e.g., local police), and equipment that can be utilized to mitigate the effects resulting from an attack on the water system. The U.S. EPA has developed ERP preparation guidance documents for small and medium systems serving populations of 3,301 to 99,999 and for large systems serving populations larger than 99,999 (U.S. EPA, 2003a; 2004a). The U.S. EPA provides an example of a water utility that underwent extensive emergency planning and an outline of the materials that should be contained in an ERP, which is referred to as a *Utility Integrated Contingency Plan* by the water utility profiled (U.S. EPA, 2008b). Key elements that should be included in ERPs are described in Table 2.5.

3.2.2. Sources of Emergency Planning Information

To assist with emergency response planning, several tools are available. The Response Protocol Toolbox contains information useful in response planning (U.S. EPA, 2003c). The Crisis Information Management Software and Field Operations and Records Management System are useful in preparing for an event, although personnel must be proficient in using these tools. Various professional and federal agencies provide guidance on emergency planning (AWWA, 2001; U.S. Army CHPPM, 2005; U.S. EPA, 2003a, 2004a). Many states also provide resources to their water utilities. The state of Washington has prepared extensive information on emergency planning for water utilities (WA DOH, 2003). The State's planning documents contain planning templates for the mission statement and goals, system information, chain of command, list of events that cause emergencies, severity of emergencies, notification list, water quality sampling list, effective communication guidelines, vulnerability assessment, response actions to specific events, alternate water sources, curtailing water use, returning to normal operation, training, and plan approval (WA DOH, 2003). As another example, the California Department of Public Health (CDPH) also provides extensive planning documentation as well as templates for ERP development (Crisologo, 2008).

In addition to assisting with planning, state agencies will likely be involved in responding to a water system threat. For example, the California

TABLE 2.5 Emergency Response Plan Components Recommended for Small/Medium and Large Water Service Providers, as Provided by the U.S. EPA

Element	Description	Small/ med.	Large
System specific information	Basic technical information that can aid water system personnel, first responders, and repair/remediation personnel during an event that may threaten water system security (e.g., contact information, population served, distribution map, process flow diagrams, engineering drawings, SCADA system operations, diagrams of chemical handling and storage facilities).	X	X
Water system roles and responsibilities	Name and contact information of a designated person to serve as the emergency response leader, who is responsible for evaluating incoming information, managing resources, coordinating with first responders, and making response decisions. An alternate emergency response leader should also be designated.	X	X
Chain-of-command chart development	Organization chart that clearly establishes individual water system staff roles and responsibilities during a security-threatening event using a well-defined command structure.		X
Communication procedures: who, what, and when	Notification lists and procedures that identify communication channels for water system staff, local entities (e.g., media, local government, public safety officials, schools), and external entities (e.g., state water regulatory agency, regional water authority, EPA, state health department).	X	X
Personnel safety	Guidance for water system staff and emergency responders on how to safely respond to threats, including facility evacuation, personnel accountability, and protective equipment.	X	X

TABLE 2.5 Emergency Response Plan Components Recommended for Small/Medium and Large Water Service Providers, as Provided by the U.S. EPA—cont'd

Element	Description	Small/ med.	Large
Identification of alternate water sources	Establishment of the quantity of water needed for various durations and identification of alternate water sources for short- and long-term water outages (e.g., emergency water shipments, emergency water supply, alternate storage and treatment sources, regional interconnections, backup wells, certified bulk water haulers).	X	X
Replacement equipment and chemical supplies	Identification of equipment that can mitigate the effect of a water system attack or threat, including current equipment, repair parts, and chemical supplies. Water systems should obtain or establish mutual aid agreements with other water utilities, the equipment and supplies needed to respond effectively to vulnerabilities identified in their VA.	X	X
Property protection	Procedures for securing and protecting water service facilities, equipment, and records during an event, including "lockdown," access control, security perimeters, evidence protection, and securing buildings against forced entry.	X	X
Water sampling and monitoring	Identification of water sampling and monitoring protocol for determining if a drinking water source is safe for public use, including sampling procedures for different types of contaminants, obtaining sample containers, confirming laboratory capabilities, and interpreting results.	X	X
Training, exercises, and drills	Program for emergency response training to inform staff members what is expected of them during a system-threatening event. Training can take several forms; for example, orientation scenarios, tabletop workshops, and functional exercises.		X

TABLE 2.5 Emergency Response Plan Components Recommended for Small/Medium and Large Water Service Providers, as Provided by the U.S. EPA—cont'd

Element	Description	Small/ med.	Large
Assessment	Periodic evaluation, in the form of an audit, of the ERP to evaluate its effectiveness and ensure the procedures and practices developed in the ERP are being implemented.		X

Department of Public Health (CA DPH) prepares and maintains Emergency Water Quality Sampling Kits, patterned after recommendations contained in the U.S. EPA RPTB (Crisologo, 2008). The kits are stored at district offices and deployed to sites where there is a threat. The samples are then transported to a state laboratory where turnaround time for analysis is estimated as 24–48 hours. During sample analyses, water utilities are expected to enact protective measures.

3.2.3. Response Guidelines

In addition to producing ERPs, development of field-ready emergency response materials is beneficial. Response guidelines (RGs) can be used to supplement ERPs and contain field-ready summaries of the information contained within the ERPs. The RGs can be very useful for first responders during an emergency or during a potential emergency.

Water providers need to develop and document potential responses to a water contamination threat, because there typically is insufficient time during a threat event to develop rigorous plans. The plans should include strategies for conducting site evaluations, collecting potentially hazardous samples, sending samples to a laboratory for analyses, instigating operational responses such as containment, and issuing public statements containing instructions regarding use of the drinking water. In support of these potential responses, water providers should have field emergency response guides ready as well as personal protective equipment (PPE), sample test kits, and rapid on-site test equipment. Public health responses (e.g., public notices) should be prepared ahead of time and be ready to deploy.

Resources provided by CDPH include action plan templates, so that water utilities can plan what activities should occur following these emergency events (Crisologo, 2008):

- Contamination to water system (possible, credible, and confirmed stages).
- Structural damage from explosive device.

- Armed intruder.
- SCADA or other type of computer security issue.
- Chlorine release.
- Power outage.
- Natural event (flood, winter storm, hurricane or tropical storm, earthquake).
- Water supply interruption.
- Bomb threat (telephone or in person, suspicious package or letter, written threat received).

3.2.4. Communication Planning

During planning, guidance should be developed and placed into the ERP on communicating with the public and examples of messages that can be used to communicate with the public and media (WA DOH, 2003). Preprepared messages, or message mapping, can greatly assist and speed up response efforts (CA DHS, 2006; CDC, 2013; U.S. EPA, 2007a). Based on data collected from previous contamination events, communication with the public is critical for maintaining public confidence (Winston and Leventhal, 2008).

3.2.5. Chain of Command

Establishing a communication plan and the chain of command for emergency situations is important. Roles and responsibilities should be clearly defined. Government officials, including public health officials, should be included in both the planning and response efforts. Contact information and structural hierarchy should be provided in accessible documentation prepared in addition to being documented in the ERP. The chain of command establishes the roles and responsibilities that take place during an emergency. Guidance on establishing a chain of command is provided by the Incident Command System (ICS) that is part of FEMA, which is part of the National Incident Management System (NIMS). The ICS is a "model tool for command, control, and coordination of a response to a public crisis" (U.S. EPA, 2003b). The ICS is an all hazards incident management concept, and it involves a common framework used across different infrastructure sectors. The ICS can be used as a field guide for responding to contamination threats. The incident command structure includes the incident commander (IC), who has overall responsibility for the response to a threat warning. The water utility emergency response manager (WUERM) is the person in charge of the response from the water provider and manages the utility's overall response to the threat. The water utility emergency operations center manager (WUOCM) is the person in charge of the emergency operations center. The information officer (IO) is in charge of information and interacts with representatives of the media. The liaison officer (LO) interacts with other agencies involved in the threat response, such as law enforcement agencies. The laboratory point of contact (LPoC), safety officer, agency representatives, and technical specialists are other personnel involved in the event of a water contamination event.

3.3. Training for Water System Threats

Water utilities need to develop training programs with periodic exercises to ensure that their personnel are knowledgeable about security-directed protocols and have sufficient opportunity to practice procedures. A training and exercise (T&E) program is an essential component of a well-functioning water security program (U.S. EPA, 2011a). According to the recommendations of the WSWG, the response staff should be clearly identified and trained (NDWAC, 2005). Although T&E is recommended, water providers are not specifically required to practice emergency response procedures. However, directed training and practice sessions are considered to be good practices and are encouraged.

Training exercises can be divided into two types: discussion based and operations based. Discussion-based exercises include seminars, workshops, tabletop exercises, and games. Operations-based exercises include drills, field exercises, functional exercises, and full-scale exercises. Exercises need to be conducted by employees on a regular basis for these efforts to be effective. The purpose of T&E is to make sure that water utilities are able to comply with the procedures outlined in their ERPs. Lessons learned during T&E programs can be used for revising documents during the planning process, which should occur on a regular basis. Another important aspect of T&E is to track and assess improvement to verify the effectiveness of T&E. Some recommendations made by the WSWG for training include publishing the results of exercises conducted, measuring response times, quantifying coordination effectiveness during exercises, rating exercise performance, considering contingency plans, and addressing the full range of threats during exercises (NDWAC, 2005).

Currently, an abundance of information is available for training and preparing water personnel for disaster and terrorism events. It is essential that water utility personnel have the opportunity to test their knowledge and apply the guidelines to their individual water systems. As an example, training exercises for water utility personnel have been facilitated by the U.S. Army Center for Health Promotion and Preventative Medicine, which can serve as a model for water providers; training scenarios used by the U.S. Army are based on actual events that occurred at military bases worldwide. Whelton et al. (2006) describe three military-style methods for training water utility personnel to respond to contamination events, which include roundtable exercises, simple functional exercises, and enhanced functional exercises. The exercises are as straightforward as gathering key personnel to discuss how a response effort would proceed, to staging an event and mimicking the response effort using teamwork and controlled communication techniques. The exercises are important for familiarizing water utility personnel with federal guidelines as well as having them meet and interact with emergency response representatives from various agencies. The practice sessions also provide personnel with experience in using the information contained within their ERPs to test out how well they can carry out the specified procedures. The exercises are particularly effective when representatives from multiple agencies participate.

Training materials for those preparing for water security events are provided by multiple agencies. The guidance provided in U.S EPA (2011a) includes templates and examples of successful T&E programs run by water providers. In addition, the U.S. EPA developed a tabletop exercise tool (TTX tool) that is consistent with the U.S. Department of Homeland Security's Homeland Security Exercise and Evaluation Program (HSEEP) (U.S. EPA, 2010a). The TTX tool is available on disk and contains 15 scenarios that can be used in exercises to allow water providers to test out their ERPs. Training materials are also available through the WaterISAC and the National Incident Management System (NIMS), which is part of the Federal Emergency Management Administration (FEMA). The National Emergency Training Center, also housed within FEMA, offers emergency response training courses through its Emergency Management Institute. Resources for emergency preparedness and response are available through other programs, such as the National Response Framework, National Disaster Recovery Framework, and the Water/Wastewater Agency Response Network. The National Association of City and County Health Officials also developed tabletop exercises that water providers can use to practice their emergency response preparedness. The standard AWWA G440-11 *Emergency Preparedness Practices* also contains emergency planning guidance.

The state of Washington provides a tabletop exercise planning guide that (1) outlines the roles and responsibilities for organizers, participants, and observers; (2) defines the exercise scope; (3) proposes a schedule; (4) provides forms to assist in carrying out exercises, including checklists, feedback forms, and message forms; (5) offers guidance on developing scenario narratives; (6) provides examples of emergency scenarios; and (7) suggests postexercise assessment activities, including reviews and corrective action plans (WA DOH, 2005). The state of Washington has also been successful in operating symposiums to engage water utilities in a discussion about water security. The Washington State Department of Health, in partnership with U.S. EPA, hosted a series of symposiums with the purpose of encouraging a security culture within the water sector, shifting thinking of goals and responsibility of water utilities and promoting an all-hazards approach. The participants in the symposiums were surveyed and responded positively.

As another example, California Department of Public Health conducts training exercises through the California Specialized Training Institute to train first responders (Crisologo, 2008). An all-hazards approach to training and emergency preparation is taken (e.g., emergencies due to earthquakes are considered). The training sessions specifically address emergency risk communication.

3.4. Planning to Respond to Wastewater System Threats

Water security is a major issue in the water industry and security is also an issue for wastewater agencies. Many of the issues relevant in water security are also pertinent to the wastewater industry. In addition, in the event of a water threat, responders may need to make decisions about discharges to the wastewater collection system.

Guidance on security has also been provided to the wastewater industry, and many of the guidelines are parallel. In 2011, the U.S. EPA developed the Wastewater Response Protocol Toolbox (WWRPTB) to assist utilities, government agencies, and emergency responders in planning for and responding to contamination threats and incidents specific to wastewater (Table 2.6). Similar to the RPTB, the WWRPTB is designed to be a tool and is not legally binding.

CONCLUSIONS

Although security has always been a concern for U.S. policy makers, the events of September 11, 2001, catalyzed a series of Executive Orders, Presidential Decision Directives, Homeland Security Presidential Directives, Presidential Policy Directives, and other legislation to facilitate government collaboration concerning major infrastructure, including water systems, in order to develop emergency response plans and strengthen security. Under these policies, the U.S. EPA became the lead federal agency for the security of the nation's water systems.

Regulating and planning for water security represents an extensive problem, as water systems span large geographic areas, consist of multiple physical components that are reliant on computer systems, have multiple access points, and are operated by both public and private agencies. The U.S. EPA is active in research and development of water security issues and has published multiple guidance tools and documents to assist water utilities in strengthening security; however, because of the variability in water systems and the fact that each system requires a unique solution, adoption of these improvements is on a voluntary basis. Therefore, the U.S. EPA focuses on vulnerability assessments and emergency response plans for contamination events, rather than mandating specific water security improvements. Available guidance tools, such as the Response Protocol Toolbox, Vulnerability Self-Assessment Tool, and Water Contaminant Information Tool, are vital resources that water utilities can rely on for better preparedness in case of contamination events.

Water utilities typically develop rigorous VAs and ERPs so that they are prepared for all types of hazards. The Bioterrorism Act required water utilities serving more than 3300 people to prepare VAs, characterizing points of attack, consequences, risk analysis, and existing countermeasures. The VAs should be updated periodically to reflect emerging threats and new tools for dealing with threats, such as system maintenance plans, risk assessment frameworks, and distribution system modeling. The ERPs contain key information and system-specific procedures, such as chain of command, personnel safety and training, communication protocols, and monitoring, needed for emergencies and contamination events.

TABLE 2.6 U.S. EPA Wastewater Response Protocol Toolbox Modules (U.S. EPA, 2011b)

Component	Title	Overview
Module 1	Wastewater Utility Planning Guide	Provides an overall guide to utility planning for contamination threats and events related to wastewater systems and includes discussion of the nature of contamination events and the activities a utility can undertake to prepare for such incidents.
Module 2	Contamination Threat Management Guide	Provides a framework for making decisions based on information available to utilities in response to a contamination threat. Included is a description of the type of information that may be useful when conducting a threat evaluation and the actions that can be implemented in response to a threat.
Module 3	Site Characterization and Sampling Guide	Describes the process of collecting information from the site of a suspected contamination event to confirm if a contamination incident did occur and whether or not it is credible. This module includes procedures for carrying out site investigation activities, such as field safety screening, field testing, and sample collection.
Module 4	Analytical Guide	Provides a framework to developing an approach for the analysis of samples collected from the site of a suspected contamination event. This module is designed to be a planning tool for laboratories and highlights lab considerations for processing emergency wastewater samples that are suspected to be contaminated.
Module 5	Public Health and Environmental Impact Response Guide	Examines the role of utilities, health officials, and regulatory agencies during a public health and environmental impact response. Includes actions that should be taken before and after a threat occurs.
Module 6	Remediation and Recovery Guide	Describes the planning and implementation of the remediation and recovery activities required to restore the water system prior to resuming normal operation after a contamination event. This module is targeted toward utility personnel, regulatory personnel, and public health entities.

Providing secure water systems is a huge undertaking for U.S. water providers, and these emergency plans may reveal that significant investments are needed to improve security and reduce risk. Therefore, water providers must be pro-active and develop efficient strategies for improving water security, such as implementing dual-use technology that benefits both day-to-day operations and increases water security. Water providers need to use available resources, aggressively plan, train personnel, and use a defensible, science-based approach to determine vulnerabilities and make sound decisions regarding water security.

REFERENCES

American Water Works Association (AWWA), 2001. Emergency planning for water utilities, AWWA Manual M19, Fourth ed. AWWA, Denver, CO.

American Water Works Association (AWWA), 2010. Risk analysis and management for critical asset protection (RAMCAP) standard for risk and resilience management of water and wastewater systems using the ASME-ITI RAMCAP plus technology, ANSI/ASME-ITI/AWWA J100-10. AWWA, Denver, CO.

Antley, A., Mapp, L., 2010. ERLN/WLA launch. Water Laboratory Alliance (WLA) Security Summit, San Francisco, CA.

American Society of Mechanical Engineers (ASME) Innovative Technologies Institute LLC (ASME-ITI), 2006. RAMCAP™: The framework, version 2.0. ASME, Washington, DC.

Biringer, B.E., Vugrin, E.D., Warren, D.E., 2013. Critical infrastructure system security and resiliency. CRC Press, Taylor and Francis Group, Boca Raton, FL.

California Department of Health Services (CA DHS), 2006. Crisis and emergency risk communication workbook for community water systems. CA DHS, Sacramento, CA.

Centers for Disease Control and Prevention (CDC), 2013. Drinking Water Advisory Communication Toolbox. Public Health Service, U.S. Department of Health and Human Services, American Water Works Association. Atlanta, GA.

Copeland, C., 2010. Terrorism and security issues facing the water infrastructure sector. Congressional Research Service, Washington, DC.

Crisologo, J., 2008. California implements water security and emergency preparedness, response, and recovery initiatives. J. Am. Water Works Assoc. 100 (7), 30–34.

DeNileon, G.P., 2001. The who, what, why, and how of counterterrorism issues. J. Am. Water Works Assoc. 93 (5), 78–85.

Haas, C.N., 2002. The role of risk analysis in understanding bioterrorism. Risk Anal. 22 (4), 671–677.

Haimes, Y.Y., 2002. Strategic responses to risks of terrorism to water resources. J. Water Resour. Plann. Manag.- ASCE 128 (6), 383–389.

Jalba, D., Cromar, N., Pollard, S., Charrois, J., Bradshaw, R., Hrudey, S., 2010. Safe drinking water: Critical components of effective inter-agency relationships. Environ. Int. 36, 51–59.

Jeong, H.J., Hwang, W.T., Kim, E.H., Han, M.H., 2009. Radiological risk assessment for an urban area: Focusing on a drinking water contamination. Ann. Nucl. Energy 36 (9), 1313–1318.

Jesperson, K., Winter 2002. Are water systems a terrorist target? On Tap Magazine. National Drinking Water Clearinghouse, Morgentown, WV.

Khan, A.S., Swerdlow, D.L., Juranek, D.D., 2001. Precautions against biological and chemical terrorism directed at food and water supplies. Public Health Rep. 116 (1), 3–14.

Linville, T.J., Thompson, K.A., 2006. Protecting the security of our nation's water systems: Challenges and successes. J. Am. Water Works Assoc. 98 (3), 234–241.

Magnuson, M.L., Allgeier, S.C., Koch, B., De Leon, R., Hunsinger, R., 2005. Responding to water contamination threats. Environ. Sci. Technol. 39 (7), 153A.

Matalas, N.C., 2005. Acts of nature and potential acts of terrorists: Contrast relative to water resource systems. J. Water Resour. Plann. Manag. 131 (2), 79–80.

Morley, K.M., 2010. Advancing the culture of security and preparedness in the water sector. J. Am. Water Works Assoc. 102 (6), 34–37.

National Drinking Water Advisory Council (NDWAC), 2005. Recommendations of the National Drinking Water Advisory Council to the U.S. Environmental Protection Agency on water security practices, incentives, and measures. National Drinking Water Advisory Council, Environmental Protection Agency, Washington, DC.

Nuzzo, J.B., 2006. The biological threat to US water supplies: Toward a national water security policy. Biosecurity Bioterrorism-Biodefense Strategy Pract. Sci. 4 (2), 147–159.

Roberson, J.A., 2011. What's next after 40 years of drinking water regulations? Environ. Sci. Technol. 45, 154–160.

Sekheta, M.A.F., Sahtout, A.H., Sekheta, F.N., Pantovic, N., Al Omari, A.T., 2006. Terrorist threats to food and water supplies and the role of HACCP implementation as one of the major effective and preventative measures. Internet J. Food Safety 8, 30–34.

Skolicki, Z., Arciszewski, T., Houck, M.H., De Jong, K., 2008. Co-evolution of terrorist and security scenarios for water distribution systems. Adv. Eng. Software 39, 801–811.

Tidwell, V.C., Cooper, J.A., Silva, C.J., 2005. Threat assessment of water supply systems using Markov latent effects modeling. J. Water Resour. Plann. Manag.- ASCE 131 (3), 218–227.

U.S. Army Center for Health Promotion and Preventive Medicine, 2005. Emergency response planning for military water systems, USACHPPM TG 297. Water Supply Management Program, Aberdeen Proving Ground, MD.

U.S. Environmental Protection Agency (U.S. EPA), 2002a. Guidance for water utility response, recovery, and remediation actions for man-made and/or technological emergencies, EPA 810-R-02-001. EPA, Washington, DC.

U.S. Environmental Protection Agency (U.S. EPA), 2002b. Vulnerability assessment factsheet, EPA 816-F-02-025. EPA, Washington, DC.

U.S. Environmental Protection Agency (U.S. EPA), 2003a. Large water system emergency response plan outline: Guidance to assist community water systems in complying with the Bioterrorism Act, EPA 810-F-03-007. EPA, Washington, DC.

U.S. Environmental Protection Agency (U.S. EPA), 2003b. Module 1: Water utility planning guide, EPA-817-D-03–001. Response Protocol Toolbox (RPTB) interim final: Planning for and responding to contamination threats to drinking water systems. EPA, Washington, DC.

U.S. Environmental Protection Agency (U.S. EPA), 2003c. Planning for and responding to drinking water contamination treats and incidents: Overview and application, EPA-817-D-03–007. Response Protocol Toolbox (RPTB) interim final: Planning for and responding to contamination threats to drinking water systems. EPA, Washington, DC.

U.S. Environmental Protection Agency (U.S. EPA), 2004a. Emergency response plan guidance for small and medium community water systems, EPA 816-R-04-002. EPA, Washington, DC.

U.S. Environmental Protection Agency (U.S. EPA), 2004b. EPA's role in water security research: The water security research and technical support action plan, EPA/600/R-04/037. EPA, Washington, DC.

U.S. Environmental Protection Agency (U.S. EPA), 2004c. Response Protocol Toolbox: Planning for and responding to drinking water contamination threats and incidents, response guidelines. EPA, Washington, DC.

U.S. Environmental Protection Agency (U.S. EPA), 2006. Water security handbook: Planning for and responding to drinking water contamination threats and incidents, EPA 817-B-06-001. EPA, Washington, DC.

U.S. Environmental Protection Agency (U.S. EPA), 2007a. Effective risk and crisis communication during water security emergencies: Report of EPA sponsored message mapping workshops, EPA-600-R-0-027. EPA, Washington, DC.

U.S. Environmental Protection Agency (U.S. EPA), 2007b. Water Contaminant Information Tool, EPA 817-F-07-001. EPA, Washington, DC.

U.S. Environmental Protection Agency (U.S. EPA), 2007c. Water security initiative: Interim guidance on planning for contamination warning system deployment, EPA 817-R-07-002. EPA, Washington, DC.

U.S. Environmental Protection Agency (U.S. EPA), 2008a. 2008 annual update to the water sector-specific plan, EPA817-K-08-002. EPA, Washington, DC.

U.S. Environmental Protection Agency (U.S. EPA), 2008b. Decontamination and recovery planning—Water and wastewater utility case study, EPA 817-F-08-004. EPA, Washington, DC.

U.S. Environmental Protection Agency (U.S. EPA), 2008c. Sampling guidance for unknown contaminants in drinking water, EPA 817-R-08-003. EPA, Washington, DC.

U.S. Environmental Protection Agency (U.S. EPA), 2008d. Water security initiative: Cincinnati pilot post-implementation system status, EPA 817-R-08-004. EPA, Washington, DC.

U.S. Environmental Protection Agency (U.S. EPA), 2008e. Water security initiative: Interim guidance on developing an operational strategy for contamination warning systems. EPA, EPA 817-R-08-002. EPA, Washington, DC.

U.S. Environmental Protection Agency (U.S. EPA), 2008f. Water security initiative: Interim guidance on developing consequence management plans for drinking water utilities, EPA 817-R-08-001. EPA, Washington, DC.

U.S. Environmental Protection Agency (U.S. EPA), 2009. National primary drinking water regulations, EPA 819-F-09-004. EPA, Washington, DC.

U.S. Environmental Protection Agency (U.S. EPA), 2010a. Tabletop exercise tool for water systems: Emergency preparedness, response, and climate resiliency. EPA 817-F-006. EPA, Washington, DC.

U.S. Environmental Protection Agency (U.S. EPA), 2010b. Vulnerability Self-Assessment Tool (VSAT), EPA 817-F-10-015. EPA, Washington, DC.

U.S. Environmental Protection Agency (U.S. EPA), 2010c. The water laboratory alliance—Response plan, EPA 817-R-10-002. EPA, Washington, DC.

U.S. Environmental Protection Agency (U.S. EPA), 2011a. How to develop a multi-year training and exercise (T&E) plan, EPA 816-K-11-003. EPA, Washington, DC.

U.S. Environmental Protection Agency (U.S. EPA), 2011b. Wastewater Response Protocol Toolbox (WWRPTB): Planning for and responding to wastewater contamination threats and incidents, EPA 817-B-09-001. EPA, Washington, DC.

U.S. Environmental Protection Agency (U.S. EPA), 2012a. Secondary drinking water regulations: Guidance for nuisance chemicals, EPA 816-F-10-079. EPA, Washington, DC.

U.S. Environmental Protection Agency (U.S. EPA), 2012b. Water Laboratory Alliance Tool Kit, EPA 817-B-12-001. EPA, Washington, DC.

U.S. General Accounting Office (U.S. GAO), 2003. Drinking water: Experts' views on how future federal funding can best be spent to improve security, Report to the Committee on Environment and Public Works, U.S. Senate, GAO-04–29. U.S. GAO, Washington, DC.

U.S. General Accounting Office (U.S. GAO), 2004. Drinking water: Experts' views on how federal funding can best be spent to improve security, Testimony Before the Subcommittee on Environment and Hazardous Materials, Committee on Energy and Commerce, House of Representatives, GAO-04–1098T. U.S. GAO, Washington, DC.

Washington State Department of Health (WA DOH), 2003. Emergency response planning guide for public drinking water systems, DOH PUB. 331–211. Environmental Health Programs, Division of Drinking Water, Olympia, WA.

Washington State Department of Health (WA DOH), 2005. Tabletop exercise planning guide for public drinking water systems, DOH PUB. #331–279. Environmental Health Programs, Division of Drinking Water, Olympia, WA.

Whelton, A.J., Wisniewski, P.K., States, S., Birkmire, S.E., Brown, M.K., 2006. Lessons learned from drinking water disaster and terrorism excercises. J. Am. Water Works Assoc. 98 (8), 63–73.

Winston, G., Leventhal, A., 2008. Unintentional drinking-water contamination events of unknown origin: Surrogate for terrorism preparedness. J. Water Health 6, 11–19.

Threats

1. INTRODUCTION

Water distribution systems are inherently vulnerable to attack, given the extensive and public nature of these systems from the source water to the consumer. The vulnerability of a water system is a function of the physical attributes of the system, relative locations of system components, and characteristics of the population relying on the water. Threats to water security include physical attacks, cyber attacks, damage due to natural disasters and accidents, and deliberate contamination caused by vandals and terrorists. Potential contaminants include biological, chemical, and radiological constituents that can result in illness or death in the exposed population. The potential contaminants represent a broad range of constituents that have different characteristics in terms of water solubility, environmental persistence, human health effects, potential for being produced in large quantities, and fate in treatment processes. Identifying and characterizing threats is a necessary step in designing appropriate prevention measures, warning systems, response protocols, and potential rehabilitation alternatives. This chapter covers water system vulnerability, potential attacks, potential contaminants, public health effects, and case studies of water systems that experienced a water system contamination event.

2. WATER SYSTEM VULNERABILITIES

In the U.S., public water systems (PWSs) are defined as "at least 15 service connections or serves an average of at least 25 people for at least 60 days a year"; as such, there are approximately 155,000 PWSs in the United States, serving 312 million people (U.S. EPA, 2011). Water distributed through the water systems is used in many ways including direct consumption, food preparation, sanitation, irrigation, medical procedures, industrial activities, recreation activities, and fire suppression.

The four portions of a water system that are vulnerable to attack are the source water, treatment system, distribution system, and other supporting systems. Within each of these portions, many components are vulnerable (Table 3.1).

Water systems may have all or only a portion of the system components listed in Table 3.1, depending on the nature of the water system. Source water varies among systems, some water systems rely solely on groundwater while others rely on surface water sources. The type of source water affects the types of components used in the water system. The physical attributes of water systems are highly variable and depend on the geographic location of the community, population density, nonresidential development, and management of the water system. Conveyance systems and treatment strategies are also highly variable. In some water systems (e.g., many found in California), water may travel long distances and undergo extensive treatment prior to being delivered to consumers. In other systems, groundwater may be pumped a short distance to consumers as a result of wells located throughout the service area. The number of pump stations, reservoirs, and the total length of piping is largely a function of geography and urban development (e.g., densely populated versus rural areas). In flat areas, it is more common to use water towers, whereas buried reservoirs are more typical in areas that have variable terrain. The sizes of water system components are highly variable; reservoirs can contain between 3 and

TABLE 3.1 Water System Components

Portion of Water System	Components
Source water	Lakes, rivers, impoundments, dams, canals, intake structures, pump stations, diversion structures, groundwater, wells
Treatment system	Chemical and holding tanks, screens, filters, disinfection facilities, other unit treatment processes, pump stations
Distribution system	Storage tanks, reservoirs, pipelines, pump stations, booster chlorination facilities, water towers, fire hydrants, air relief valves, backflow preventers
Other	Conveyance facilities, personnel, computer systems, control centers, electrical systems, communications systems, laboratories, chemical supply, maintenance facilities

30 million gallons (Nuzzo, 2006). The diversity of systems requires that each system or service area must develop an independent approach to water security.

Water treatment technologies can vary depending on the quality of the source water. Groundwater may undergo very little treatment prior to being conveyed to customers. For example, only softening or iron removal may be necessary prior to chlorination. Surface water and groundwater under the influence of surface water are typically filtered and disinfected according to federal drinking water regulations. However, a few unfiltered surface waters remain in the United States. Unfiltered surface water sources are vulnerable because fewer barriers remove potential contaminants and the water supply can be contaminated at any location between the water source and the end user. Unfiltered surface water sources are located in remote watersheds with variable security. In some cases, access to the watershed is strictly controlled; and in other cases public access is allowed. The Bull Run Watershed (Portland, Oregon) is restricted by federal law to human entry. The Cedar River Municipal Watershed (Seattle, Washington) is restricted to public access and the Quabbin Reservoir (Boston, Massachusetts) allows public access but is highly regulated. For all water systems, source water protection is an important aspect of maintaining high-quality water and the task of protecting the water source can take many forms.

The vulnerability of individual components is related to location, ease of access, number of downstream water users, and the like. Protecting distribution systems is problematic because of the large, diffuse nature of the systems, public access (through the plumbing systems located inside buildings), and the fact that distribution systems occur downstream of treatment facilities. Distribution systems contain components specifically designed to permit access, such as fire hydrants and valve vaults. In densely populated areas, it can be difficult to secure the water system, because the residence times in the system are short and vulnerability is a function of the large population served. In general, water systems in the vicinity of a more populated city are more susceptible to attack because there are more targets and a contamination event would be more extreme. Source water facilities may be located in remote areas that are difficult to secure without considerable effort, staff, and resources. Some conveyance structures extend long distances, such as the California Aqueduct (over 700 miles), making it difficult to protect such a geographically extensive water system component.

3. THREATS TO WATER SYSTEMS

To understand the vulnerability of water systems (Table 3.1), a multitude of potential methods for attacking water systems should be considered. Attacks on a water system fall under one of the following three categories:

- Physical attack.
- Cyber attack.
- Deliberate contamination.

Physical attacks threaten the main system components (e.g., dam, treatment plant, reservoir, pipelines, electrical and communication systems), which may eliminate an essential part of the water system. Taking essential water system components out of service may not necessarily lead to death and health problems in the population. However, disruption of water service may have implications for disrupting civil order. Cyber attacks may be carried out by infiltrating the supervisory control and data acquisition (SCADA) system or managing operations of a water system and stopping or increasing the addition of chlorine, fluoridation, or otherwise changing operations. The outfall from cyber attacks may or may not result in health problems in the served population. Deliberate contamination could occur by injection of biological agents, biotoxins, or chemicals (including radiological chemicals) into the water system. Given that the possibility of causing harm to the population is greater in the case of deliberate contamination, there is a significant probability that a terrorist would use this method to attack a water system. The remainder of this book focuses on deliberate and accidental contamination events.

The effect of contamination events depends on the extent of dilution, contaminant characteristics (toxicity for chemicals and virulence for biological agents), environmental fate (persistence and decay), routes of exposure (e.g., ingestion, dermal contact), and host susceptibility (WHO, 2002). The U.S. EPA Response Protocols Toolbox (RPTB) Module 1 contains a discussion on the possibility of water contamination threats versus the probability of these threats (U.S. EPA, 2003a). While some contaminants could cause massive health effects, most potential contaminants would cause only local effects, although it is likely a contamination event would cause panic even if it did not result in a significant health effect (U.S. EPA, 2003a). As an example, injecting a harmless dye into the water system could cause fear and panic in the affected population, even in the absence of health effects. An examination of some potential contaminants can give a better understanding of the potential effects of deliberate, malicious attacks.

Water supply systems are designed to protect the customer from natural biological contamination, and the same systems have potential efficacy against deliberate biological and chemical contamination. Khan et al. (2001) lists five factors that are protective against biological contamination events:

1. Dilution.
2. Disinfection.
3. Natural attenuation from hydrolysis, sunlight, and biological processes.
4. Filtration.
5. Limited exposure from drinking water (1–1.5 liters daily, depending on region).

With the exception of disinfection, these factors are also protective against chemical contaminants. Drinking water systems are designed to protect customers against microbial contamination by maintaining a disinfectant residual in the distribution system. Of course, the disinfectant residual present may be insufficient to be protective against all biological contaminants. Protecting a water system against chemical threats is more difficult. However, disinfection could also prove protective if there is a reaction between the chemical contaminant and the disinfectant.

For contaminants that are toxic or infectious at low concentrations, it can be wrongfully assumed by water providers that only a small quantity of contaminant would be necessary to have a large impact or affect a large population. However, dispersal of the contaminant could be difficult because water systems are inherently not well mixed. For example, pipe flow follows a "plug flow" pattern with little dispersion. Reservoirs are somewhat well mixed; however, short-circuiting does occur. Additionally, a water distribution system comprises portions with short detention times and high turnover as well as portions with relatively stagnant water. For water systems that have sources (e.g., wells) throughout the service area, it may not be possible to effectively contaminate a large portion of the system, because of poor mixing within the pressurized distribution system. Dispersing a contaminant to a large population would require introduction of the contaminant nearer to the source water, which may mean upstream of treatment processes. However, upstream contamination of a water system results in a large degree of dilution. Essentially, dispersing a contaminant throughout a water system to produce a significant impact and affect a large population is not simple.

4. POTENTIAL CONTAMINANTS

Potential contaminants that could be used in deliberate attacks include biological, chemical, and radiological agents. To be a concern in water systems, these agents must be:

- Water soluble.
- Stable in environmental systems.
- Toxic or infectious in low doses.
- Capable of being produced or acquired in large quantities.
- Not affected by treatment processes or the disinfection residuals found in distribution systems.

In addition to these characteristics, contaminants of concern that are odorless and tasteless are more difficult to detect and therefore may be more dangerous. However, immediate panic and disruption may be less with tasteless and odorless compounds. Contaminants that do not alter the color or add particulates to the water are also more difficult to detect. The extent to which contaminants are soluble, stable, and toxic varies. In addition, the ability to remove and inactivate contaminants in treatment systems varies as well.

Many potential biological and chemical warfare agents are available on a worldwide basis, although the U.S. and other governments have established agreements, such as the Chemical Weapons Convention (CWC) and Biological Weapons Convention (BWC), to limit the growth of this malicious industry.

Chemical weapons, as defined by the CWC, are identified in the Schedule 1 list. In the classified report commissioned by the U.S. Department of Homeland Security, potential threats and mitigation strategies are reviewed, and it was determined that "There are more than 60,000 commercially available chemical and biological agents and approximately 760 radionuclides that are naturally occurring or used in commerce or research" (Porco, 2010). After screening the list of potential contaminants, 21 were identified as high-priority agents, which pose the most threat for contaminating a water system; however, this list is included in a report that is considered "For Official Use Only" and is not reproducible (Porco, 2010).

There are several sources of information for identifying potential threats. The password-protected Water Contaminat Information Tool (WCIT) contains a database of potential contaminants (U.S. EPA, 2007). The water threat notification organization WaterISAC also provides secure information about potential threats; this information is available only to authorized entities, such as water utilities. Other sources of information include the CDC website, the U.S. EPA RPTB, and reports from the World Health Organization (WHO) (CDC, 2013; U.S. EPA, 2003a, 2003b; WHO, 2004). The U.S. EPA contaminant candidate list (CCL) is another source of contaminant information, although not all of the agents on the list are suitable for use in an attack. Contaminants on the CCL are those that could be present in drinking water and have, or are suspected to have, health effects. The Safe Drinking Water Act Amendments (1996) require the CCL to be published by the U.S. EPA every five years to identify future contaminants for regulation (Richardson, 2009). Many of the chemicals on the CCL are industrial compounds. There are 104 chemical contaminants and 12 microbiological contaminants on the third CCL that has been issued (Appendix C). Richardson (2009) provides a discussion of emerging contaminants, many of which are on the CCL.

4.1. Biological Contaminants

Biological contaminants consist of pathogenic microorganisms and biotoxins that can cause illness and death in the exposed and infected population. Biological contaminants can be bioterrorism agents if they can be produced in mass quantities in forms that are capable of causing disease in the exposed populations (CDC, 2013). When used in biological warfare, these bioterrorism agents become weapons of mass destruction (WHO, 2004). Not all bioterrorism agents can be used in an attack on a drinking water system; only waterborne agents that are environmentally stable can be feasibly used.

It is important to understand the characteristics of biological contaminants and the diseases they cause to better prepare for water threats and to clarify the potential repercussions. In addition to understanding the potential repercussions of various biological contaminants, identification of potential biological threats is important for understanding the potential of treatment processes to remove and inactivate these agents. The treatability and fate of many potential biological contaminants in environmental systems is unclear (Burrows and Renner, 1999). Some high water threat pathogens, such as *Vibrio cholerae, Salmonella,* and *Shigella,* can be inactivated relatively easily using chlorine. Anthrax, plague-causing agents, and *Cryptosporidium,* on the other hand require alternative treatment and disinfection methods (Burrows and Renner, 1999). Multiple barriers, a succession of treatment processes, are most effective for removal of biological contaminants (Nuzzo, 2006).

4.1.1. Pathogens

Pathogenic microorganisms consist of viruses, bacteria, protozoa, and helminthes that can cause infection in the exposed population. A minimum dose is required to cause infection. Not all infections produce symptoms, although the infections can lead to disease and even death. Pathogens infect hosts to replicate and affect the health of the host. Infectivity is described using dose-response models and the median infectious dose (ID-50) that causes infection in half of the exposed population (Burrows and Renner, 1999). Likewise, the probabilistic relationship between morbidity and mortality is unique to each pathogen. The ID-50 can vary greatly, even among very similar pathogens, although the reasons for these differences are still unclear (Leggett et al., 2012). The susceptibility of a host to infection is a function of multiple factors and some portions of the population, such as children, the elderly, and immune-compromised individuals, are known to be more susceptible than others. In addition, individuals can gain immunity from repeated exposures to pathogens. Information for dose-response models for many pathogens is available in the published literature although some researchers have stated that the data is incomplete or has not yet been adequately applied (Haas, 2002; Nuzzo, 2006).

In general, a high concentration of bacteria is required to cause infection by contaminated drinking water, making it difficult to contaminate a large water system. The infectious dose for protozoa, helminthes, and viruses is much lower than for most bacteria, making them more likely to be used as a biological contaminant.

4.1.2. Biotoxins

The human health effects of exposure to biotoxins vary from those resulting from infection by pathogenic microorganisms. Biotoxins are chemical substances produced by bacteria, plants, and fungi that serve as a biological defense and poison, which protects the toxin-producer from being consumed. In this case there is an exposure-response model, and a median lethal dose (LD-50) is used to describe the health effects. Immunity to chemical toxins does not occur similar to immunity to pathogens; however, some portions of the population are more affected than others to similar exposure. One way to describe toxicity is to use a relative ranking that is the water solubility of a contaminant divided by the lethal dose, with botulinum toxin being the most toxic of the compounds analyzed (Clark and Deininger, 2000). Human health data for biotoxins is mostly unavailable, so rodent models are used instead (Burrows and Renner, 1999). Biotoxins are of particular concern as environmental contaminants because they are toxic at extremely low doses (Arnon et al., 2001; Paterson, 2006). Biotoxins are also odorless, colorless, and tasteless (Khan et al., 2001).

4.1.3. Examples of Biological Contaminants

Bioterrorism agents can be classified according to their ability to cause harm to an exposed population and how relatively easy or difficult it is to weaponize these agents (Table 3.2) (CDC, 2013). These agents can cause disease through exposure to contaminated air, water, or food. According to the CDC classification system, Category A bioterrorism agents represent the highest risk while Category C agents are considered emerging threats (CDC, 2013).

Input for the classification system and biological contaminant selection was given during a meeting of experts that convened in 1999 (Rotz et al., 2002). The group used the following criteria to select biological contaminants of concern in bioterrorism: (1) public health impact, (2) potential of agent to be weaponizable, (3) public perception, and (4) public health preparedness requirements (Rotz et al., 2002). Formulation of the classification system and list were based on use of a risk-based matrix. The CDC maintains a list of potential bioterrorism agents although not all of the agents may be feasible for contaminating water systems (Table 3.3).

The U.S. EPA RPTB contains a comprehensive list of contaminants that could be used in an intentional contamination event and provides several examples of

TABLE 3.2 Centers for Disease Control and Prevention Classification of Bioterrorism Agents (CDC, 2013)

Category	Traits
Category A	Easily disseminated or transmitted from person to person Result in high mortality rates and have the potential for major public health effects Might cause public panic and social disruption Require special action for public health preparedness
Category B	Moderately easy to disseminate Result in moderate morbidity rates and low mortality rates Require specific enhancements of CDC's diagnostic capacity and surveillance
Category C	Available Ease of production and dissemination Potential for high morbidity and mortality rates and major health impact

each type of contaminant. Twenty pathogens and five biotoxins are on the U.S. EPA RPTB list (Table 3.4).

In a comprehensive review of bioterrorism agents, Burrows and Renner (1999) listed possible bioterrorism agents, with information on health effects, potential for use as weapons, environmental stability, and treatability of potential contaminants. Burrows and Renner (1999) listed 18 pathogens and 9 biotoxins that could be used to intentionally contaminate water systems (Table 3.5). The biotoxins listed were aflatoxin, anatoxin, botulinum toxins, microcystins, ricin, saxitoxin, Staphylococcal enterotoxins, T-2 mycotoxin, and tetrodotoxin (Burrows and Renner, 1999).

It is not clear that all the threats identified can be weaponized or easily deployed. Rotz et al. (2002) identified the organisms responsible for anthrax, smallpox, plague, botulinum toxin, and tularemia as the most probable biological

TABLE 3.3 Bioterrorism Agents and Diseases Based on Centers for Disease Control and Prevention Classification (CDC, 2013)

Category	Microorganism	Disease
Category A	*Bacillus anthracis*	Anthrax
	Clostridium botulinum toxin	Botulism
	Yersinia pestis	Plague
	Variola major	Smallpox
	Francisella tularensis	Tularemia
	Filoviruses (e.g., Ebola, Marburg)	Viral hemorrhagic
	Arenaviruses (e.g., Lassa, Machupo)	fevers
Category B	*Brucella* species	Brucellosis
	Epsilon toxin of *Clostridium perfringens*	
	Food safety threats (e.g., *Salmonella* species, *Escherichia coli* O157:H7, *Shigella*)	
	Burkholderia mallei	Glanders
	Burkholderia pseudomallei	Melioidosis
	Chlamydia psittaci	Psittacosis
	Coxiella burnetii	Q fever
	Ricin toxin from *Ricinus communis* (castor beans)	
	Staphylococcal enterotoxin B	
	Rickettsia prowazekii	Typhus fever
	Alphaviruses (e.g., Venezuelan equine encephalitis, eastern equine encephalitis, western equine encephalitis)	Viral encephalitis
	Water safety threats (e.g., *Vibrio cholerae*, *Cryptosporidium parvum*)	
Category C	Emerging infectious diseases, such as Nipah virus and hantavirus	

TABLE 3.4 Biological Contaminants Listed in the U.S. EPA Response Protocol Toolbox (U.S. EPA, 2003a)

Category	Subcategory	Examples
Microbiological	Bacteria	*Bacillus anthracis, Brucella* spp., *Burkholderia* spp., *Campylobacter* spp., *Clostridium perfringens, E. coli* O157:H7, *Francisella tularensis, Salmonella* typhi, *Shigella* spp., *Vibrio cholerae, Yersinia pestis, Yersinia enterocolitica*
	Viruses	Caliciviruses, enteroviruses, Hepatitis A/E, Variola, VEE virus
	Parasites	*Cryptosporidium parvum, Entamoeba histolytica, Toxoplasma gondii*
Biotoxins		Ricin, saxitoxin, botulinum toxins, T-2 mycotoxins, microcystins

TABLE 3.5 Diseases List by Level of Threat (Burrows and Renner, 1999)

Low Water Threat	Medium Water Threat	High Water Threat
Glanders	Brucellosis	Anthrax
Melioidosis	*Clostridium perfringens*	Cholera
Typhus	Psittacosis	Plague
Encephalomyelitis	Q fever	Salmonellosis
Hemorrhagic fever	Smallpox	Shigellosis
		Tularemia
		Enteric viruses
		Cryptosporidiosis

weapons. Since biotoxins are toxic at extremely low concentrations, they would be more feasibly used. Botulinum toxins are thought to have human health effects at the extremely low dose of 0.0004 µg/L (Burrows and Renner, 1999) and are thought to represent an extreme threat as a biological weapon (Arnon et al., 2001).

4.2. Chemical Contaminants

A lot of chemicals could be used in an attack. The potential for a chemical agent to be a weapon of mass destruction (WMD) depends on the toxicity and dose-response relationship. Chemical agents that could be used to contaminate a water system fall under several categories: volatile organic chemicals (e.g., chloroform), semi-volatile organic chemicals (e.g., organophosphates such as malathion), and nonvolatile organic chemical (e.g., surfactants). Additional chemical agents that could be used include pharmaceuticals, metals, arsenic salts, and cyanides. The U.S. EPA RPTB contains a detailed list of potential chemical agents (Table 3.6) (U.S. EPA, 2003a). Many of the potential contaminants in the RPTB list are readily available to the public and often available in large quantities. Examples of readily available chemicals include pesticides, herbicides, paint thinners, and toilet bowl cleaners. While it is possible that a terrorist would be able to obtain restricted biological and chemical agents, it is more likely that common chemicals would be used.

The list in Table 3.6 represents chemicals that are known to have a high probability of being used in an attack. However, the list of potential chemical agents is much larger, and it is estimated that thousands of new chemicals are introduced into the market each year. The cumulative amount of toxic chemicals produced in the United States is not measured, but production of toxic chemical in the European Union was estimated to be 218 million metric tons per annum (http://epp.eurostat.ec.europa.eu). It is likely that U.S. production is equivalent or greater, suggesting that bulk quantities of toxic chemical are readily available. For example, in 2009, over 600 million pounds of pesticides were sold in California alone (http://www.cdpr.ca.gov). Many of these pesticides are potentially toxic or harmful to humans.

The contaminants of concern in drinking water are household products (personal care products and cosmetics), pharmaceuticals, commercialized manufacturing chemicals, herbicides, and pesticides. The CCL is a good source of information for chemicals that could be an issue in water systems (Appendix C). Although the CCL can be useful for identifying potential contaminants, the contaminants of concern during times of normal operations are very different from the contaminants that could be present as the result of an intentional contamination event. For example, the CCL contains ingredients in pharmaceuticals and personal care products, which may not be the most likely choice of contaminant for a terrorist.

Clark and Deininger (2000) identify the following chemicals as being possibly used as warfare agents and are extremely toxic: VX, O-ethyl-S-(N3N-dimethylaminoethyl) methyl thiophosphonate, sarin, nicotine, colchicine, cyanide, amiton, fluroroethanol, sodium fluoroacetate, selenite, arsenite, and arsenate.

4.2.1. Classifying Chemical Contaminants

A variety of indicators can be used to classify chemical contaminants. These indicators include acute toxicity, median lethal dose, production and acquisition capabilities, solubility in water, environmental stability, and treatability.

Acute Toxicity and LD-50

LD-50 is the median lethal dose of a substance, or the amount required to kill 50% of a given test population. It is a measurement used in toxicology studies

TABLE 3.6 Chemical Contaminants Listed in the U.S. EPA Response Protocol Toolbox (U.S. EPA, 2003a, p. 14–15)

Category	Subcategory	Examples
Inorganic chemicals	Corrosives and caustics	Toilet bowl cleaners (hydrochloric acid), tree-root dissolver (sulfuric acid), drain cleaner (sodium hydroxide)
	Cyanide salts or cyanogenics	Sodium cyanide, potassium cyanide, amygdalin, cyanogen chloride, ferricyanide salts
	Metals	Mercury, lead, osmium, their salts, organic compounds, and complexes (even those of iron, cobalt, and copper are toxic at high doses)
	Nonmetal	Oxyanions, organononmetals Arsenate, arsenite, selenite salts, organoarsenic, organoselenium compounds
Organic chemicals	Fluorinated organics	Sodium trifluoroacetate (rat poison), fluoroalcohols, fluorinated surfactants
	Hydrocarbons and their oxygenated or halogenated derivatives	Paint thinners, gasoline, kerosene, ketones (e.g., methyl isobutyl ketone), alcohols (e.g., methanol), ethers (e.g., methyl tert-butyl ether, MTBE), halohydrocarbons (e.g., dichloromethane, tetrachloroethene)
	Insecticides	Organophosphates (e.g., malathion), chlorinated organics (e.g., DDT), carbamates (e.g., Aldicarb) some alkaloids (e.g., nicotine)
	Malodorous, noxious, foul-tasting, and lachrymatory chemicals	Thiols (e.g., mercaptoacetic acid, mercaptoethanol), amines (e.g., cadaverine, putrescine), inorganic esters (e.g., trimethylphosphite, dimethylsulfate, acrolein)
	Organics, water miscible	Acetone, methanol, ethylene glycol (antifreeze), phenols, detergents
	Pesticides other than insecticides	Herbicides (e.g., chlorophenoxy or atrazine derivatives), rodenticides (e.g., superwarfarins, zinc phosphide, α- naphthyl thiourea)
	Pharmaceuticals	Cardiac glycosides, some alkaloids (e.g., vincristine), antineoplastic chemotherapies (e.g., aminopterin), anticoagulants (e.g., warfarin). Includes illicit drugs such as LSD, PCP, and heroin.
Schedule 1 chemical warfare agents	Schedule 1 chemical weapons	Organophosphate nerve agents (e.g., sarin, tabun, VX), vesicants, [nitrogen and sulfur mustards (chlorinated alkyl amines and thioethers, respectively)], Lewisite

to determine the potential effect of toxic substances on different types of organisms based on different routes of exposure (e.g., dermal, oral). Lower values of LD-50 are regarded as more toxic, as it means a smaller amount of the toxin is required to cause death. The globally harmonized system of classification and labeling of chemicals assigns toxic substances into five categories of acute toxicity exposure based on their LD-50, as presented in Table 3.7, with Category 1 being the most severe toxicity category and Category 5 containing chemicals that can be a hazard to certain populations but are typically of low acute toxicity.

Production and Acquisition

It is more likely that chemicals with limited acquisition regulations will be used to intentionally contaminate a drinking water system. While there is no central source of U.S. chemical production data, a few assumptions can be made to help identify chemicals that can be purchased with relative ease. For example, chemicals used as insecticides, pesticides, and herbicides are produced in large quantities for agricultural industries, and some of these chemicals can be purchased with relative ease by consumers. On the other hand, VX, a nerve gas developed for chemical warfare, is highly regulated. The development, production, stockpiling, and transfer of VX and other chemical weapons are banned under the Chemical Weapons Convention international treaty.

Water Solubility and Environmental Stability

For a chemical to be used successfully in a water system incursion, it must be soluble, to some degree, in water at temperatures typically observed in drinking water systems. Chemicals that have low solubility can still be used as contaminants; however, more of the chemical is required to reach LD-50 concentrations. Likewise, a chemical will not be a viable water contaminant if it is unstable during transport or in aquatic conditions.

TABLE 3.7 Globally Harmonized System of Classification and Labeling of Chemicals for Acute Toxicity from Oral Exposure

Relative Toxicity	Category	LD-50 Range (mg/kg)
Extremely toxic	1	≤5
	2	5–50
	3	50–300
	4	300–2000
Hazardous	5	2000–5000

Treatability

As shown in Table 3.8, conventional water treatment processes are not generally targeted at removing chemicals; rather, they are aimed to remove sediment, pollutants associated with sediment, and microorganisms. A study conducted by Westerhoff et al. (2005) conclude that conventional treatment processes can achieve some chemical removal, but specific chemical constituents are difficult to target and removal is sporadic. Other water treatment processes such as air stripping, adsorption, and ion exchange are targeted at chemical removal, and in some cases (e.g., ion exchange), pollutant removal can be selective; however, these processes are not widely used.

4.2.2. Chemical Classification of CCL-3

The contaminant criteria matrix is a useful tool to predict the likelihood of contaminants being used in an intentional contamination event. The classification procedure was applied to the U.S. EPA list of unregulated chemical contaminants (Appendix C). The list contains several chemicals that have no readily available information for animal studies and acute toxicity, which means they

---

TABLE 3.8 Water Treatment Processes

Water Treatment Process	Purpose
Conventional processes	
Screening	Debris removal
Coagulation/flocculation	Form floc and large particles
Sedimentation	Particle/floc settling and removal
Filtration	Remove algae, sediment, clay, organic/inorganic particles, and microorganisms Reduce turbidity Reduce membrane clogging
Disinfection	Inactivate microorganisms
Other processes	
Air stripping	Remove volatile chemicals
Adsorption	Remove taste and odor compounds, synthetic organic compounds, disinfection by-product precursors, and inorganic chemicals
Advanced oxidation	Degrade and potentially remove organic chemicals, effective for disinfection
Ion exchange	Remove dissolved ionic constituents

are not suspected to be toxic to humans and can be screened out (Table C.1). In addition, multiple chemicals that are insoluble in water have an LD-50 of over 5000 mg/kg or cannot achieve toxic threshold concentration in 1 liter, which is a reasonable consumption level, can be screened from the list. Based on the previous criteria, the list was cut from 104 unregulated contaminants to 29 potential threats (Table C.2). These contaminants can be further screened by considering the volume of the substance required to contaminate a water supply. For example, as shown on Table C.1, it would take nearly 960 m^3 of ethylene glycol to contaminate a 1 million gallon water supply to a concentration such that a 60 kg person drinking 1 liter of water would reach the LD-50 amount. This equates to 4395 standard shipping drums (55 gallons) of the substance needed to be introduced into the water supply.

4.3. Radiological Contaminants

Radiological contamination comes from radionuclides in the form of alpha, beta, or gamma radiation. Examples of radionuclides are radon, radium, and uranium (Khan et al., 2001). Radionuclides are typically highly regulated and secure; however, there are many sources of radionuclides throughout the world and not all are guarded with the same security standards (Gonzalez, 2003). Additionally, there are many "orphan" sources that have been abandoned or are unsecured (Gonzalez, 2003). Threats from radionuclide contaminants pose some of the most significant threats to human health; and while the effects of exposure can be experienced within hours, long-term health effects can also result from lower doses (Porco, 2010). The U.S. EPA RPTB identifies potential radiological contaminants, based on their wide use in medical applications and relative availability (Table 3.9). Some radionuclides are not a concern as they are not available in large quantities or have short half-lives (Paperiello, 2003).

Cesium-137 was used in medical applications for the treatment of cancers. Radioactive cesium in the form of cesium chloride, a powdery salt, is highly soluble in water and is a contaminant of concern (Gonzalez, 2003). Colbalt-60, also used in radiation therapy, is usually in the form of solid metal pellets and, due to limited water solubility, is not easy to disperse. However, because typical

TABLE 3.9 Radiological Contaminants Listed in the U.S. EPA Response Protocol Toolbox (U.S. EPA, 2003a)

Radiological Contaminants	Half-Life
Cesium-137	30.2 years
Iridium-192	73.8 days
Cobalt-60	5.3 years
Strontium-90	28.8 years

radiation therapy cobalt capsules can contain up to 1000 pellets and soluble cobalt salts can be synthesized, dispersion is still possible (Gonzalez, 2003). Iridium-192, used for industrial radiography, is also in metal pellet form but can exist in a water soluble form. Iridium-192 may be easier to steal than other radionuclides due to its industrial setting; however, it is difficult to accumulate large quantities, as sources are stored across separate industries (Gonzalez, 2003). Strontium-90, found in large quantities in radioisotope thermogenerators, is of concern, as many thermogenerators were abandoned during the dissolution of the Soviet Union and new ruling governments did not act to secure them (Gonzalez, 2003). Strontium-90 is also widely used in radiotherapy to treat cancer and strontium salts can be water soluble.

5. PUBLIC HEALTH INFORMATION

In addition to understanding the potential contaminants that could be used in an attack, information on the resulting health effects for the exposed population should also be collected. Information on the possible symptoms that could occur following exposure to biological and chemical warfare is available through the CDC and the WHO (CDC, 2013; WHO, 2004). For example, symptoms of infection, such as diarrhea, vomiting, and stomach cramps, are suggestive of infection by biological agents (CDC, 2013). In addition, Meinhardt (2005) describes the associated threats potential contaminants pose from the public health perspective. Similar to the U.S. EPA's role in securing water systems from attack, the CDC is responsible for addressing the pubic health aspects of national security. To fulfill their leadership position in public health, the CDC publishes information on potential health effects and monitors health care data to detect outbreaks and exposure events to determine the cause. Using disease surveillance systems to detect waterborne disease outbreaks is problematic because many infected people do not seek medical care (Khan et al., 2001). Fortunately, biological attacks are uncommon (Ashford et al., 2003).

Exposure to contaminants from potable water can occur through multiple pathways. Water distributed in the water systems is used in many ways, including direct consumption, food preparation, sanitation, irrigation, medical procedures, industrial activities, recreation activities, and fire suppression. Exposure to contaminants in potable water may be due to ingestion, dermal contact, or the inhalation of volatile chemicals. Ingestion occurs when the water is consumed directly or as the result of consuming food prepared using water that had been contaminated. Dermal contact occurs as the result of bathing, swimming, hand washing, and cleaning. Inhalation can occur as the result of showering and exposure to irrigation systems. The method of exposure (e.g., inhalation or ingestion) for the same contaminant can lead to different symptoms, and thus, the same contaminant may have different LD-50 values for each method of exposure (Burrows and Renner, 1999; CDC, 2013).

Another aspect of inherent variability for potential water contamination events is that members of the exposed population respond differently to threats and exhibit different health effects. To describe the variability in the health response of individuals, probability models are used to describe the relationship between the dose and the most probable response (Calabrese and Baldwin, 2003). Data are collected through animal studies, outbreaks, and epidemiological studies. Dose-response models for consuming contaminated drinking water have also been adapted to work with water system network software and public health exposure information to predict health risks and assist in developing management strategies to reduce risk (Murray et al., 2006).

Dose-response models are similar for exposure to biological and chemical agents, but there are distinct differences. The human health endpoints may also be different. For pathogens, the commonly used endpoints are infection and death; however, immunity must also be considered. For chemicals, infection is not relevant and the cumulative lifetime exposure of chemicals should be considered. Dose-response relationships are often described using the median lethal dose (LD-50) or the median infectious dose (ID-50). To apply the dose-response models to determine the probability of a health endpoint such as death, the concentration of the contaminant must be determined as well as the quantity of contaminated water consumed. For chemical exposure models, the weight of the exposed individual must also be determined to ascertain the impact of exposure. Information regarding typical quantities of water consumed and body weight is available, as summarized in Table 3.10 (U.S. EPA, 2004).

TABLE 3.10 Typical Daily Water Consumption by Body Weight (U.S. EPA, 2004)

Age Category	Mean Weight (kg)		Mean Water Consumption (ml/kg/day)	
	Male	Female	Male	Female
<0.5 years	6	6	47	54
0.5–0.9	9	9	35	47
1–3	14	14	21	21
4–6	21	20	21	20
7–10	32	31	15	14
11–14	52	50	13	11
15–19	73	61	12	11
20+	84	68	14	16

The threat of the contaminant depends on the infectious dose and the symptoms caused by the contaminant. Biological agents with a lower infectious dose and more extreme symptoms exhibit a greater threat. These agents can also spread throughout the population more easily through means such as secondary transmission. Biological agents to which the general population has little or no immunity are more dangerous.

Exposure to high amounts or radiation can lead to either acute radiation syndrome, or chronic radiation syndrome. Acute radiation syndrome is typically expressed within 24 hours of exposure, and symptoms can include nausea, vomiting, diarrhea, headache, fever, cognitive impairment, fatigue, loss of hair, reduced white blood cell count, hemorrhage, shock, and death. Chronic radiation syndrome is usually expressed months or years after exposure and typically results in cancer. The severity of radiation exposure depends on the method of exposure, with ingestion usually being more severe due to uptake and incorporation of radionuclides into cells.

6. CASE STUDIES

A water security breach can occur as the result of major disasters, emergencies, terrorist attacks, terrorist threats, civil unrest, wilderness or urban fires, floods, hazardous material spills, nuclear accidents, earthquakes, hurricanes, tornadoes, tropical storms, tsunamis, war-related disasters, public health and medical emergencies, and other occurrences requiring an emergency response (CIPAC, 2009). To understand the potential effects of a water security breach, the potential problems that could occur as the result of an intentional intrusion or contamination event, and the appropriate responses to these events, it is prudent to study past incidences related to water system failures that include accidents, outbreaks, and natural disasters.

Water contamination events are rare but do occur occasionally in localized cases. Beering (2002) presents a brief history of the use of biological agents to carry out malevolent acts. Gray (2008) conducted an investigation of water contamination events in the United Kingdom and found that many small-scale events do occur. Ashford et al. (2003) reviewed information from outbreak investigations worldwide that were recorded by the CDC from 1988 to 1999 and found that, in 44 of the 1099 investigations, the causative agent had bioterrorism potential (*V. cholerae*, *Y. pestis*, viral hemorrhagic fever, *B. anthracis*, *C. botulinum* toxin, *F. tularensis*, *R. rickettsiae*). However, bioterrorism or intentional contamination appeared possible in only 6 of the 1099 outbreak investigations (Ashford et al., 2003). Gleick (2006) provides a history of terrorist threats and attacks on water systems and infrastructure, many of which were not carried out. In addition to attacks on water systems, there is a history of biological agents being used in warfare and in various malicious attacks on humans (Christopher et al., 1997).

Camelford, England, 1988

An unintentional water contamination event occurred at the Lowermoor water treatment facility, which supplied drinking water to 20,000 people at the time of the event. Twenty tons of aluminum sulfate was accidentally dumped into a hopper directly connected to the drinking water supply. The public was not warned, and when they asked about the change in texture and acidic taste, the public was assured that the water was safe to drink (Winston and Leventhal, 2008). This instance resulted in an aluminum concentration of 620 mg/L, which is more than 3000 times the recommended concentration by both the U.S. EPA and the World Health Organization. The system was flushed with potable water to correct this situation and returned to normal system operation. However, a loss of public confidence ensued following the event. The UK Department of Health determined health risks unlikely; however, debate continues regarding the long-term effects of aluminum ingestion on the residents' health (Jalba et al., 2010)

Charlotte, North Carolina, 1997

Firefighters inadvertently caused chemical foam agents to enter the drinking water system, which resulted in "do not drink" and "do not use" notifications for residents for several days (Baecher, 2006).

Los Angeles, California, 2000

Firefighting foam was accidentally injected into the water system at a fire hydrant during a firefighting response (Mehta, 2000). Residents in 700 households were warned not to use the water for any purpose and the system was flushed prior to being placed back into service. Exposure to high concentrations of the foam, which contained several chemicals, such as ethyl alcohol and ethylene glycol, could have caused mucus-membrane and skin irritation. However, the foam was released in extremely low concentrations and was not expected to cause detrimental health effects to the residents.

Milwaukee, Wisconsin, 1993

One of the most significant waterborne disease outbreaks that has occurred in modern times in the United States was the *Cryptosporidium parvum* outbreak in Milwaukee, Wisconsin. There was deterioration in the source (Lake Michigan) raw-water quality due to spring rains, runoff from farmlands, and sewage overflows. This caused a decrease in effectiveness of the filtration system and resulted in high turbidity of the treated water. As a result of the outbreak, over 400,000 people became ill, approximately 4,000 people were hospitalized, and an estimated 54 people died. Oocysts were identified by analyzing ice made from the water during the outbreak (Mackenzie et al., 1994). The estimated cost due to illnesses and productivity losses was $96 million (Corso et al., 2003).

New Orleans, Louisiana, 2005

Hurricane Katrina offers lessons learned in responding to a large-scale disaster situation, which is helpful in preparing for a potential water system contamination event (Whelton et al., 2006). After Hurricane Katrina, thousands of people were forced to relocate to a common area, which left people susceptible to communicable diseases. The lack of accessibility to basic needs, such as clean water, food, and health care, increased the emotional stress and probability of spreading diseases (Brodie et al., 2006). It has been reported that Katrina evacuees were exposed to water-related diseases, including norovirus, *Salmonella*, and toxigenic and nontoxigenic *V. cholerae*, which have been attributed to the lack of safe water (Watson et al., 2007).

Røros, Norway, 2007

An outbreak caused by the bacterium *Campylobacter* appeared to have been linked with consumption of contaminated drinking water (Jakopanec et al., 2008). A medical officer in Røros reported 15 patients with gastroenteritis caused by the *Camplyobacter* on May 7, 2007, and by May 10, 2007, he estimated that hundreds of people had become infected. The probable cause of the outbreak was pressure reduction in the distribution system that resulted in a cross connection. It was not possible to definitively attribute the outbreak to the potable water system.

Tel Aviv, Israel, 2001

Ammonia was accidentally improperly dosed in the water system (Winston and Leventhal, 2008). The public was promptly notified and advised not to consume the water, even if boiled. Following flushing of the affected water system, operation returned to normal.

Walkerton, Ontario, Canada, 2000

Following heavy spring rains, a municipal water system well appeared to have become contaminated with *E. coli* O157:H7 and *Campylobacter jejuni* originating from cow manure, causing 2,300 illnesses and seven deaths (Hrudey et al., 2003).

CONCLUSIONS

Public water systems are inherently vulnerable to attack, given the extensive and public nature of these systems. Depending on the nature of the water system, vulnerability of system components can vary greatly, as vulnerability depends on the individual system components, location, ease of access, and number of downstream water users. To understand the vulnerability of water systems, physical attacks, cyber attacks, and deliberate contamination events should be considered.

Potential contaminants that could be used in a deliberate attack include biological, chemical, and radiological agents. The greatest threats to water systems are contaminants that are water soluble, environmentally stable, highly toxic or infectious, available in large quantities, tasteless and odorless, and not affected by treatment processes. Because water supply systems are designed to reduce natural biological contamination through dilution, disinfection, natural attenuation, and filtration, these same systems may reduce deliberate biological and chemical contamination. However, there are fewer protections against contaminants introduced downstream of treatment facilities.

Information for identifying potential contamination threats is widely available to water professionals. The password-protected WCIT, U.S. EPA RPTB and a review by Burrows and Renner (1999) contain comprehensive lists of contaminants that could be used. Fortunately, international agreements, such as the Chemical Weapons Convention and Biological Weapons Convention, limit the production and access to the most dangerous potential contaminants. Although it is unclear which contamination threats are easiest to weaponize or deploy, tools such as the contaminant criteria matrix can predict the likelihood of contaminants being used in a contamination event.

Understanding the health effects resulting from potential contaminants is vital during contamination events and should be included in response planning. Symptoms from exposure to biological and chemical agents are available through the CDC and the WHO; however, the effect on the exposed population can vary based on demographics and the pathway of exposure. To assist in estimating health effects, dose-response models have been developed for exposure to both biological and chemical agents, where the threat depends on the dose and the symptoms.

Additionally, the physical, chemical, and biological properties of potential contaminants are valuable for conducting sampling and analyses as well as developing the most appropriate response protocols. Contaminant properties can also be used in the rehabilitation of contaminated water systems.

REFERENCES

Arnon, S.S., Schechter, R., Inglesby, T.V., Henderson, D.A., Bartlett, J.G., Ascher, M.S., Eitzen, E., Fine, A.D., Hauer, J., Layton, M., Lillibridge, S., Osterholm, M.T., O'Toole, T., Parker, G., Perl, T.M., Russell, P.K., Swerdlow, D.L., Tonat, K. Working Grp Civilian, B, 2001. Botulinum toxin as a biological weapon—Medical and public health management. JAMA-J. Am. Med. Assoc. 285 (8), 1059–1070.

Ashford, D.A., Kaiser, R.M., Bales, M.E., Shutt, K., Patrawalla, A., McShan, A., Tappero, J.W., Perkins, B.A., Dannenberg, A.L., 2003. Planning against biological terrorism: Lessons from outbreak investigations. Emerg. Infect. Dis. 9 (5), 515–519.

Baecher, G.B., 2006. Mitigating water supply system vulnerabilities. Protection of civilian infrastructure from acts of terrorism. In: Frolov, K.V., Baecher, G.B. (Eds). Springer, Dordrecht, The Netherlands.

Beering, P.S., 2002. Threats on tap: Understanding the terrorist threat to water. J. Water Resour. Plann. Manag.- ASCE 128 (3), 163–167.

Brodie, M., Weltzien, E., Altman, D., Blendon, R.J., Benson, J.M., 2006. Experiences of Hurricane Katrina evacuees in Houston shelters: Implications for future planning. Am. J. Public Health 96 (8), 1402–1408.

Burrows, W.D., Renner, S.E., 1999. Biological warfare agents as threats to potable water. Environ. Health Perspect. 107 (12), 975.

Calabrese, E.J., Baldwin, L.A., 2003. The Hormetic dose-response model is more common than the threshold model in toxicology. Toxicol. Sci. 71 (2), 246–250.

Centers for Disease Control and Prevention (CDC), 2013. Emergency preparedness and response: Preparing for AND responding to specific hazards. Centers for Disease Control and Prevention, Atlanta, GA.

Christopher, G.W., Cieslak, T.J., Pavlin, J.A., Eitzen, E.M., 1997. Biological warfar, a historical perspective. JAMA-J. Am. Med. Assoc. 278 (5), 412–417.

Clark, R.M., Deininger, R.A., 2000. Protecting the nation's critical infrastructure: The vulnerability of U.S. water supply systems. J. Contingencies Crisis Manag. 8 (2), 73–80.

Corso, P.S., Kramer, M.H., Blair, K.A., Addiss, D.G., Davis, J.P., Haddix, A.C., 2003. Cost of illness in the 1993 waterborne *Cryptosporidium* outbreak, Milwaukee, Wisconsin. Emerg. Infect. Dis. 9 (4), 426–431.

Critical Infrastructure Partnership Advisory Council (CIPAC) Working Group, 2009. All-hazard consequence management planning for the water sector: Preparedness, emergency response, and recovery. Water Environment Federation (WEF), Alexandria, VA.

Gleick, P.H., 2006. Water and terrorism. Water Policy 8, 481–503.

Gonzalez, A., 2003. Security of radioactive sources: Threats and answers. International Conference on Security of Radioactive Sources. International Atomic Energy Agency, Vienna, Austria.

Gray, J., 2008. Water contamination events in UK drinking-water supply systems. J. Water Health 6, 21–26.

Haas, C.N., 2002. The role of risk analysis in understanding bioterrorism. Risk Anal. 22 (4), 671–677.

Hrudey, S.E., Payment, P., Huck, P.M., Gillham, R.W., Hrudey, E.J., 2003. A fatal waterborne disease epidemic in Walkerton, Ontario: Comparison with other waterborne outbreaks in the developed world. Water Sci. Technol. 47 (3), 7–14.

Jakopanec, I., Borgen, K., Vold, L., Lund, H., Forseth, T., Hannula, R., Nygard, K., 2008. A large waterborne outbreak of campylobacteriosis in Norway: The need to focus on distribution system safety. Bmc. Infect. Dis. 8 (128).

Jalba, D., Cromar, N., Pollard, S., Charrois, J., Bradshaw, R., Hrudey, S., 2010. Safe drinking water: Critical components of effective inter-agency relationships. Environ. Int. 36, 51–59.

Khan, A.S., Swerdlow, D.L., Juranek, D.D., 2001. Precautions against biological and chemical terrorism directed at food and water supplies. Public Health Rep. 116 (1), 3–14.

Leggett, H.C., Cornwallis, C.K., West, S.A., 2012. Mechanisms of pathogenesis, infective dose and virulence in human parasites. PLoS Pathogens 8, e1002512.

Mackenzie, W.R., Hoxie, N.J., Proctor, M.E., Gradus, M.S., Blair, K.A., Peterson, D.E., Kazmierczak, J.J., Addiss, D.G., Fox, K.R., Rose, J.B., Davis, J.P., 1994. A Massive outbreak in Milwaukee of *Cryptosporidium* infection transmitted through the public water-supply. N. Engl. J. Med. 331 (3), 161–167.

Mehta, S., 2000. Firefighting foam taints cities' water. Los Angeles [CA] Times. May 23, 2000.

Meinhardt, P.L., 2005. Water and bioterrorism: Preparing for the potential threat to US water supplies and public health. Annu. Rev. Public Health 26, 213–237.

Murray, R., Uber, J., Janke, R., 2006. Model for estimating acute health impacts from consumption of contaminated drinking water. J. Water Resour. Plann. Manag- ASCE 132 (4), 293–299.

Nuzzo, J.B., 2006. The biological threat to US water supplies: Toward a national water security policy. Biosecurity Bioterrorism-Biodefense Strategy Pract. Sci. 4 (2), 147–159.

Paperiello, C., 2003. Identification of high priority radioactive sources based on risk. International Conference on Security of Radioactive Sources. International Atomic Energy Agency, Vienna, Austria.

Paterson, R.R.M., 2006. Fungi and fungal toxins as weapons. Mycol. Res. 110, 1003–1010.

Porco, J.W., 2010. Municipal water distribution system security study: Recommendations for science and technology investments. J. Am. Water Works Assoc. 102 (4), 30–32.

Richardson, S.D., 2009. Water analysis: Emerging contaminants and current issues. Anal. Chem. 81 (12), 4645.

Rotz, L.D., Khan, A.S., Lillibridge, S.R., Ostroff, S.M., Hughes, J.M., 2002. Public health assessment of potential biological terrorism agents. Emerging Infect. Dis. 8 (2).

U.S. Environmental Protection Agency (U.S. EPA), 2003a. Module 1: Water utility planning guide, EPA-817-D-03–001. Response Protocol Toolbox (RPTB) interim final: Planning for and responding to contamination threats to drinking water systems. EPA, Washington, DC.

U.S. Environmental Protection Agency (U.S. EPA), 2003b. Module 3: Site characterization and sampling guide, EPA-817-D-03–003. Response Protocol Toolbox (RPTB) interim final: Planning for and responding to contamination threats to drinking water systems. EPA, Washington, DC.

U.S. Environmental Protection Agency (U.S. EPA), 2003c. Module 4: Analytical guide, EPA-817-D-03–004. Response Protocol Toolbox (RPTB) interim final: Planning for and responding to contamination threats to drinking water systems. EPA, Washington, DC.

U.S. Environmental Protection Agency (U.S. EPA), 2004. Estimated per capita water ingestion and body weight in the United States–An update, EPA-822-R-00-001. EPA, Washington, DC.

U.S. Environmental Protection Agency (U.S. EPA), 2007. Water Contaminant Information Tool, EPA 817-F-07-001. EPA, Washington, DC.

U.S. Environmental Protection Agency (U.S. EPA), 2011. Public drinking water systems: Facts and figures. EPA, Washington, DC.

Watson, J.T., Gayer, M., Connolly, M.A., 2007. Epidemics after natural disasters. Emerging Infect. Dis. 12 (1).

Westerhoff, P., Yoon, Y., Snyder, S., Wert, E., 2005. Fate of endocrine-disruptor, pharmaceutical, and personal care product chemicals during simulated drinking water treatment processes. Environ. Sci. Technol. 39 (17), 6649–6663.

Whelton, A.J., Wisniewski, P.K., States, S., Birkmire, S.E., Brown, M.K., 2006. Lessons learned from drinking water disaster and terrorism excercises. J. Am. Water Works Assoc. 98 (8), 63–73.

Winston, G., Leventhal, A., 2008. Unintentional drinking-water contamination events of unknown origin: surrogate for terrorism preparedness. J. Water Health 6, 11–19.

World Health Organization (WHO), 2002. Environmental health in emergencies and disasters: A practical guide. WHO, Geneva, Switzerland.

World Health Organization (WHO), 2004. Public health response to biological and chemical weapons, WHO guidance. WHO, Geneva, Switzerland.

Chapter 4

Prevention

Chapter Outline

1. INTRODUCTION

Prevention is one of the first barriers in a defense-in-depth approach to water security. Maintaining physical barriers, such as alarms, fences, and locks, reduces the likelihood of threats to water systems. Other effective methods to improve water security include employee screening and training. The term *water system hardening* is used to describe the enhancement of physical security of a water system to reduce the vulnerability of the water system to attacks. The technologies described in this chapter are intended to make water systems less vulnerable and reduce the risk of a water contamination event.

2. MOTIVATION FOR PHYSICAL PROTECTION PLANNING

There are multiple reasons to provide physical protection infrastructure at water system facilities and at the locations of water system components. In addition to protecting the water supply and quality, protection of water system facilities protects valuable property against multiple types of attacks. *Design basis threat* (DBT) is a term used in the security industry to define adversaries and their capabilities. In regard to water systems, DBTs can consist of vandals that wish to damage property, criminals that intend to steal property, saboteurs that intend to cause damage and are "typically motivated by political, doctrinal, or religious causes," and insiders that intend to cause damage (ASCE et al., 2006). Insiders may be disgruntled employees or people working with criminals or saboteurs. The DBTs may be trespassing on water utility property with the intent to damage or steal property or they may be attempting to interrupt water service or contaminate the water supply.

3. WATER SYSTEM COMPONENTS REQUIRING PROTECTION

All aspects of the water supply, treatment, and distribution system are vulnerable to attack, necessitating protection at all levels. Categories of components of the water system requiring protection include raw water facilities (e.g., impounds, intake structures), wells and pumping stations, water treatment plants (including basins, pump stations, clear wells), finished water storage facilities, distribution systems elements (valves, hydrants, backflow prevention valves, air release valves), and water system support facilities (chemical storage, fuel storage, electrical and communication infrastructure) (ASCE et al., 2006). Note that many structures include air vents and overflow piping, which allow access into key facilities, such as wellheads and clear wells. Access to vents and piping should be restricted as part of a physical protection plan.

Determining which aspects of the water supply require protection depends on multiple factors, including effectiveness of current security measures, probability of being targeted, and consequences of a contamination event (U.S. EPA, 2007). Tidwell et al. (2005) proposed a vulnerability assessment framework based on Markov latent effects (MLE) modeling. This probabilistic approach allowed these researchers to calculate asset security for various facilities as a function of asset characteristics and protective measures taken by water utilities.

Drinking water supplies in the United States are derived from surface water (80%) and groundwater (20%); much of the surface water is impounded in reservoirs behind dams, and these sources require protection (Baecher, 2006). In some water systems, water travels long distances from the source to the end users; en route, much of the water travels in open conveyance structures that are vulnerable to attack (Baecher, 2006).

4. WATER SYSTEM HARDENING

Water system hardening is the use of physical barriers in water systems to deter intentional tampering with water infrastructure and to deny access to facilities. Haimes (2002) describes the hardening of a water system to produce a system more robust and resistant to attack. Water hardening can take many forms. One way to improve security is to restrict entry at facilities such as water sources, wellheads, pump stations, treatment facilities, and reservoirs. Restricting entry can be accomplished using fences, walls, gates, and the like. Restricting entry may also be accomplished by locking gates and leaving doors closed if restrictive barriers are already in place. Water hardening can be introduced by adding system redundancy, so that system functionality can continue despite having some components out of service. Examples of redundancy include providing backup power supplies, equipment, and water supplies (such as standby wells that can be turned on in the event of an emergency). The resilience of a system refers to its ability to continue functioning following a failure or attack.

Physical protection systems for water systems are meant to (1) deter intruders, (2) detect intruders, (3) delay intruders, and (4) initiate a response to the intrusion (ASCE, 2006). A multicomponent system of interrelated parts is needed to provide all of these functions. One advantage with water system hardening with physical barriers is that it allows water providers to determine if their security has been compromised, which is evident by broken locks, open doors, cut barbed wire, and so forth. Alarm systems for water facilities may be more expensive than for other facilities, because of the complex requirements and necessity to interface with other remote data collection systems (Moses and Bramwell, 2002). The security systems should be expandable, and it may be necessary to provide multiple groups access to facilities and sites (e.g., public works and parks departments), making the systems more complex. In addition, alarm systems must be designed to operate in corrosive environments, such as in the vicinity of chlorine storage tanks (Moses and Bramwell, 2002).

Federal funding is not available for water hardening projects, so it is important for water utilities to plan for improvements as part of their capital facilities planning processes. Federal funding and support; however, has been provided to assist with planning and training efforts. Abundant information is available from the U.S. EPA to assist water utilities in water hardening efforts. It is important for water utilities to maximize the dual benefits, as mentioned earlier, of water security improvements and to communicate these benefits to their customers.

5. WATER SYSTEM PHYSICAL PROTECTION TECHNOLOGY AND DEVICES

Prevention barriers to enhance water security may include background checks on potential employees, alarms, fencing, lighting, locks, protection of essential documents, a badge system for entry into key facilities, and guards (Linville and Thompson, 2006). Enhanced security monitoring includes intrusion detection devices on water facilities to generate alerts that are broadcast locally or sent to a centralized control center (Allgeier et al., 2011). Physical hardening can also include locating facilities within buildings and using resistant building materials, such as blast-resistant glass, secure doors and frames, concrete and masonry in lieu of light metal framed buildings, and so forth (AWWA, 2009). Table 4.1 contains a list of equipment and devices commonly used in hardening a water system to prevent contamination events and other incursions. Vendors and products located as part of this study are listed in Appendix D. The list

TABLE 4.1 Security Features, Devices, and Equipment for Water Systems and Facilities

Security Feature	Description
Fencing and perimeter walls	Chainlink, ornamental fencing, or perimeter wall
	Anticlimb and anticut fencing
	Fencing may include topping such as barbed wire
	Intrusion detection devices on fencing
	Fence foundation enhancements to deter tunneling efforts
	A second interior fence or perimeter wall
Entrance gates	Secured with key locks, numeric keypad, or card access system
	Remote control with intercom for visitors and vendors or a staffed guardhouse
Vehicular barriers	Bollards, retractable bollards, wedge barriers, crash beams, jersey barriers, security planters
	Barriers located at site perimeter, key facilities, roads, vehicle doors
Clear zones	Vegetation-free zone around site perimeter
	Buildings designed to provide unobstructed views
	Clear areas provided throughout the site
	Landscaping that does not obscure views of buildings or facilities
Dedicated facilities for visitors and vendors	Public and visitor parking located away from key facilities
	Designated waiting area for visitors
	Protocols for receiving visitors, including sign-in procedures
	Requiring visitors to display identification at all times
	Protocols for receiving deliveries and packages
	Dedicated meeting room located outside secured areas for meetings with visitors and vendors
Personnel identification	Requiring employees to wear badges
Site lighting	Throughout site, at the perimeter, and at gate entrances
	Motion-activated lighting

Facility signs	Site entrance signs "No Trespassing" signs Signs discrete so as not to bring attention to key facilities
Site utilities	Fuel storage tanks located away from buildings and perimeter Electrical equipment, such as transformers, generators, and switchgear, in locked panels or cages Locking devices on contractor's temporary connections
Distribution system security measures	Manholes, valve vaults, valve operators, sampling stations, monitoring wells, and well casings with shrouded locks, possibly with intrusion detection Screens over large openings (e.g., grates on culverts, screens on vents) Protective grating, screens, or covers to protect open basins Protective locked cage on exposed equipment, pipelines, well casing air relief valves, and air lines extending through well casing Fire hydrants secured with locking mechanisms Elevated vault hatch lids that prevent inflow
Building security features	Exterior doors with key locks or electronic access, blast-resistant, with tamper-resistant door hinges, with intrusion alarms Interior doors to critical facilities with automatic locking and access control Building entrance with double entry system (sally port entrance) Windows with break-resistant or blast-resistant glass, possibly with glass-break detection Windows located away from doors to prevent access to doors through windows Interior motion detection and associated camera system Barriers (e.g., grills) at skylights, louvers, and roof hatches, possibly with intrusion detection Roof access ladders with shrouded locks, possibly with intrusion alarms
Backflow prevention	Flow of water limited to one direction, to prevent backflow from backpressure or backsiphonage Prevention installed at potentially hazardous cross-connections (e.g., industrial facilities) Types: air gap, double check valves, reduced pressure, vacuum breakers Designed for multifamily residences Located throughout service area

(Continued)

TABLE 4.1 Security Features, Devices, and Equipment for Water Systems and Facilities—Cont'd

Security Feature	Description
Closed circuit television (CCTV)	Fixed cameras at facility exterior doors, hatches, and vaults Pan-tilt-zoom cameras at main gate, impoundments, intakes, open channels, main entrance door, and interior protected areas
Online water quality monitoring system	See Chapter 5, "Detection"
Power and communication facilities	Power panels locked All electrical and communication wiring placed in conduit Backup power provided for security components (e.g., UPS) Redundancy provided for communication paths and critical utility connections Supervisory control and data acquisition infrastructure in locked PLC/RTU enclosures with tamperproof switches on enclosures
Chemical facilities	Chemical fill stations at building exteriors locked, possibly with intrusion detection Chemical storage and feed equipment locked, possibly with intrusion detection
Clear well facilities	Hatches with shrouded locks, possibly with intrusion detection Vents with double screens, shrouded locks, and intrusion alarm Overflow outlet with screen or flap valve in cage, possibly with intrusion detection Automated isolation valve with remote operation possible

Alarms and intrusion sensors	Perimeter sensors used on walls, fences, doors, windows
	Perimeter sensor types: foil, magnetic switches, glass-break detectors
	Intrusion sensors located throughout site and within buildings
	Interior sensors triggered when intrusion does not occur through doors or windows
	Interior sensor types: active and passive infrared, quads, ultrasonic, microwave, dual-technology, buried line, fence-mounted, linear beam, glass-break, door and hatch contact alarm switches, motion sensors, pipeline vibration detection
	Alarm displayed locally or relayed to a centralized alarm system
	Exterior intrusion sensors: buried, fence associated, freestanding
	Interior intrusion sensors: boundary-penetration sensors, interior motion sensors, proximity sensors
Equipment enclosures	Located outdoors and above ground
	Used for backflow prevention devices, valves, pumps, motors
	Types: one piece that drops over, removable top, sectional, shelters with access doors
	Connected to foundation with secure mounting brackets
Access control systems	Biometric hand, finger, and iris recognition
	Card identification systems: proximity, Wiegand, magnetic stripe, bar code, Hollerith, infrared, barium ferrite, Smartcards

Source: Adapted from ASCE et al., 2006; AWWA, 2009

represents a wide range of security features that are available to assist in water system hardening.

The security features listed in Table 4.1 represent a wide range of strategies for protecting vulnerable water systems from attack. Many of these security features (e.g., fences and gates) are likely already in place at many water facilities. Some of the security features could be easily adapted (e.g., locked doors throughout sites). Implementing various protocols at facilities is an effective way of improving water security. For example, protocols are needed that are understood and followed by all on-site personnel. Protocols are needed for chemical deliveries, package delivery, and receiving visitors. Security-directed employment procedures and methods for maintaining secure information and control systems are also needed. Integrating water security features into new facilities is one way to address deficiencies in water systems. Regular inspections of critical portions of the infrastructure are important for maintaining security.

Measures in addition to the security features listed in Table 4.1 can be taken to secure a water system. Restricting access to vital areas, such as open basins, hazardous materials sites, chemical injection sites, and the water system source water, can help prevent contamination events. The SCADA system for water system management should be updated for increased security to prevent cyber attacks. Quick closing valves should be eliminated throughout the system whenever possible, as closing these valves can cause a sudden increase in pressure that may rupture pipes and damage sensitive equipment and valves. Fire hydrants can be upgraded to be tamper resistant at the same time that corrosion resistant technology is implemented. Simple, inexpensive devices can be installed to prevent backflow, thus protecting against contamination, while having no negative effects on hydrant performance for firefighting (Smith et al., 2010).

Water providers have always attempted to secure critical facilities, and since the terrorist attack of September 11, 2001, security devices around sites have become even more commonplace. For example, water providers use fencing and gates to secure sites. Additional hardware that is used to secure sites include door security, hatch locks, locked access to ladders, manhole locks, lockout devices for valves, and vent security. Security barriers that prevent vehicle access to sites are important for preventing contamination threats. Large quantities of contaminants are necessary to significantly affect the water system, and it would not be possible to transport the necessary quantity of contaminants to the site without a vehicle. Alarms are used to notify water providers of trespassers and break-ins and can be tied into SCADA systems or telemetry systems already in use for managing the water system. Alarms have become more sophisticated and can include covert sensors that monitor movement and body heat. Equipment enclosures are additional devices that can be used to prevent access to equipment.

The need for additional technology to secure water facilities and prevent incursion has resulted in new products being developed and marketed. Detection sensors and alarm systems are becoming increasingly advanced. Wireless security systems can be more cost effective, under some scenarios, than traditional systems, because less electrical infrastructure is required (Moses and Bramwell, 2002). Determining the appropriate technology to implement is based on site-specific needs, such as environmental conditions (e.g., fog, snow, ice, corrosiveness), installation surface, proximity to buildings and pipelines, terrain, sensitivity to nearby motion and animals, maintenance requirements, sustainability, reliability, timeliness, aesthetics, and cost (NRC, 2011; U.S. EPA, 2007). More information on the installation and testing of the following security technologies can be found in the U.S. Nuclear Regulatory Commissions' report, *Intrusion Detection Systems and Subsystems* (NRC, 2011).

5.1. Exterior Sensor Systems

Multiple types of exterior sensor systems are available for securing facilities. Readily available exterior sensor technologies include microwave sensors, active infrared sensors, electric field sensors, porter coaxial cable systems, taut wire sensors, and fence disturbance sensors (NRC, 2011). All external sensors systems must be able to successfully detect intruders and any attempts to bypass or interfere with the system. Some systems, such as fence disturbance sensors, detect fence deflections and vibrations and are easy and inexpensive to implement, as most facilities already have existing boundary fencing. Taut wire and electric field sensors, on the other hand, can be adapted to work with existing walls, roofs, or fences. Other systems, such as microwave and infrared sensors, require the installation of separate sensor posts at greater cost. Sensor systems must be carefully installed to provide adequate coverage and ensure power and communication lines are secure. Sensor systems must also be thoroughly tested to determine any deficiencies in coverage and vulnerabilities due to crawling, jumping, climbing, bridging, tunneling, or tampering (NRC, 2011).

5.2. Interior Sensor Systems

Interior sensor systems are a secondary line of defense for the detection of intruders. Available interior sensor systems include balanced magnetic switches, interior microwave sensors, passive infrared, proximity sensors, dual-technology sensors, and video motion detection (NRC, 2011). Balanced magnetic switches are a cheap but effective technology for monitoring the opening and closing of doors using a magnetic switch above the door frame. Proximity sensors, usually a pressure or strain gauge, capacitor, or switch, are typically used to guard a specific asset or location and are the last line of detection systems. Dual-technology sensors use a combination of two sensor methods to reduce the chance of false alarms. Interior microwave and passive infrared sensors are used to detect movement and changes in thermal energy, respectively. Video motion detection causes video surveillance systems to increase the frame rate and signal an alarm when motion is detected. Interior sensor systems should not be used as the only means of detection but should be used in combination with exterior sensors and video systems (NRC, 2011). Like exterior sensor systems, interior systems should be installed and thoroughly tested to avoid deficiencies in coverage or operation.

5.3. Video Systems

Video systems consist of cameras and digital video recorders to record and alert operators when motion is detected. Closed circuit television (CCTV) systems allow for the real time visualization of camera outputs and the archiving of events on removable media (NRC, 2011). Additional communication networks may be required for the transmission of recorded videos for off-site analysis, making the installation of video systems very costly (U.S. EPA, 2007). Camera systems can be divided into multiple categories: black and white, color, day/night, thermal, and intensified (NRC, 2011). Color cameras provide more detail during the day, while black and white cameras work better at night under artificial lighting. Day and night cameras combine the two and can switch between color and black and white depending on ambient light conditions. Thermal imaging cameras detect thermal radiation using gradients of colors or shades of grey, and are best used at night. Image intensified cameras, or night vision cameras, are also ideal for night surveillance. It is important that ambient lighting be bright enough to illuminate an intruder, but also to avoid camera hot spots and contrast issues (NRC, 2011). Appropriate ambient lighting in itself may also act as a deterrent.

5.4. Emergency Backup

Emergency backup power systems are crucial to maintain water system security in the case of natural disasters or system tampering. These systems should be able to automatically switch on without causing any system alarms or loss in security functions (NRC, 2011). The three most common backup power supplies are uninterruptible power supplies (UPSs), engine generators, and batteries (NRC, 2011). UPSs are generally placed between the systems and the power supply, and contain a system of batteries, chargers, switches, and inverters to maintain charge while conveying power to systems. Generators are typically diesel engines and should be designed to meet the load requirements of the security system. Batteries are typically set up in parallel to the load and source to allow for float charging and to seamlessly power the system when necessary. Backup power supplies should be kept in secure facilities with intrusion detection and surveillance (NRC, 2011). They should also be properly maintained and monitored to prevent any degradation in performance.

6. WATER SYSTEM SECURITY STANDARDS AND GUIDELINES

The U.S. EPA does not produce standards for preventative water security strategies. However, it publishes Water Security Guides on its website with descriptions on physical asset monitoring and control products and provides information on specific manufacturers, contact information, and product descriptions. In addition, the American Society of Civil Engineers (ASCE), American Water Works Association (AWWA), and the Water Environment Federation (WEF) jointly worked on the Water Infrastructure Security Enhancements Project that resulted in guidance documents and recommendations for water provider physical security improvements (ASCE et al., 2006). The AWWA standard *Security Practices for Operation and Management* also provides guidance on prevention measures (AWWA, 2009).

7. CAPITAL IMPROVEMENT PLANNING AND MAINTENANCE

In addition to utilizing the security-enhancing products described in this chapter, water systems should meet all federal and state health standards. Maintaining a well-designed and well-operated water system is beneficial in ensuring a more secure system. Water distribution system standards outlined in the Ten State Standards are shown in Table 4.2. In particular, it is important for water providers to have an adequate number of valves to isolate portions of the water system and have the means to flush all portions of the system using the recommended flushing velocity of 2.5 feet per second. Note that, in the event of a water contamination event, sufficient water supplies should be available to provide water for firefighting.

TABLE 4.2 Recommendations for Water Distribution System According to the Ten State Standards (www.10statestandards.com)

Component	Standard
Minimum pressure	20 psi (140 kPa)
Normal working pressure	60 to 80 psi (410–550 kPa) and no less than 35 psi (240 kPa)
Minimum size for water lines that provide fire protection	6 inch diameter
Minimum size water main without fire protection	3 inch diameter
Flushing velocity in pipes	2.5 feet per second
Valve spacing in commercial districts	500 feet
Valve spacing in noncommercial districts	800 feet

In addition to implementing measures for water systems to prevent incursions and access to the water supply, it is possible to operate and maintain the water system in such a way as to minimize the damage that would be incurred in the event of a contamination event. Planning is an important component of preparing for a water contamination event. In addition, maintaining a regular schedule of flushing and disinfecting pipelines and holding tanks is important for minimizing the presence of biofilms and scaling in the water system. If a contamination event occurs, the rehabilitation effort will be less severe if scaling and biofilms are minimal. Selection of pipe materials may affect the extent of scaling and biofilms as well.

The devices and equipment described in this section represent large investments for water providers, regardless of the size of the agency. Fortunately, many of the devices serve dual purposes, such as ensuring safety (e.g., by providing ladder guards) and preventing vandalism (e.g., by providing multiple intrusion sensors). Many of the items discussed in this section require significant and regular maintenance, which results in increased operating costs (e.g., oversight on alarm systems). Water providers must properly plan for these costs.

CONCLUSIONS

Prevention is the most basic component of a defense-in-depth approach to water security. All aspects of the water supply, treatment, and distribution system are vulnerable to attack, necessitating protection at all levels. Maintaining physical barriers, such as alarms, fences, and locks reduces the likelihood of threats to water systems while securing valuable facility property. Other effective methods to improve water security include employee screening and training. Many security systems are already in place at many water facilities and can be easily adapted to protect vulnerable system components.

Water system hardening is the use of physical barriers to enhance the physical security of a water system and reduce intentional tampering, vandalism, theft, and attacks. Physical protection systems deter, detect, and delay intruders while providing a response to the intrusion. Determining which aspects of the water supply require protection depends on the current security measures, probability of being targeted, and consequences of a contamination event.

Water providers have always attempted to secure critical facilities, and since the terrorist attack of September 11, 2001, security devices at sites have become even more commonplace. The need for additional technology to secure water facilities and prevent incursion has resulted in innovative solutions being developed and marketed, including multiple types of interior and exterior sensor systems, emergency backup power systems, and corrosion protection for hydrants. The lack of U.S. EPA regulations and federal funding for preventative water security strategies reduces the adoption rate of modern security features. Therefore, it is important that innovative and cost-effective dual-use technologies and planning strategies be developed to compensate for the large investment required to implement such strategies.

REFERENCES

Allgeier, S.C., Haas, A.J., Pickard, B.C., 2011. Optimizing alert occurrence in the Cincinnati contamination warning system. J. Am. Water Works Assoc. 103 (10), 55–66.

American Society of Civil Engineers (ASCE), American Water Works Association (AWWA), Water Environment Federation (WEF), 2006. Guidelines for the physical security of water utilities ASCE, Reston, VA.

American Water Works Association (AWWA), 2009. Security practices for operation and management, ANSI/AWWA G430-09. AWWA, Denver, CO.

Baecher, G.B., 2006. Mitigating water supply system vulnerabilities. In: Protection of civilian infrastructure from acts of terrorism. K.V. Frolov and G.B. Baecher (Eds). Springer, Dordrecht, The Netherlands.

Haimes, Y.Y., 2002. Strategic responses to risks of terrorism to water resources. J. Water Resour. Plann. Manag.- ASCE 128 (6), 383–389.

Linville, T.J., Thompson, K.A., 2006. Protecting the security of our nation's water systems: Challenges and successes. J. Am. Water Works Assoc. 98 (3), 234–241.

Moses, J., Bramwell, M., 2002. Champlin Water Works seeks right level of security against terrorist threat. J. Am. Water Works Assoc. 94 (4), 54–56.

Smith, E.D., Ginsberg, M., VanBlaricum, V.L., Hock, V., Ehmann, A., 2010. Demonstration of a corrosion-resistant retrofit system to upgrade fire hydrants. Corrosion and Prevention 2010, Adelaide, Australia.

Tidwell, V.C., Cooper, J.A., Silva, C.J., 2005. Threat assessment of water supply systems using Markov latent effects modeling. J. Water Resour. Plann. Manag.- ASCE 131 (3), 218–227.

U.S. Environmental Protection Agency (U.S. EPA), 2007. Water security initiative: Interim guidance on planning for contamination warning system deployment, EPA 817-R-07-002. EPA, Washington, DC.

U. S. Nuclear Regulatory Commission (NRC), 2011. Intrusion detection systems and subsystems, technical information for NRC licensees, NUREG-1959. Office of Nuclear Security and Incident Response, NRC, Rockville, MD.

Detection

1. INTRODUCTION

Detection of contaminants in drinking water systems is a critical component of a defense-in-depth approach to water security. If preventative measures fail and an incursion occurs, detection of the contamination event is necessary to develop a swift response and a sensible rehabilitation plan. Real-time detection provides the basis for quick response, which minimizes the amount of contaminated water delivered to consumers and therefore reduces exposure. Real-time detection of contaminants is important because it is difficult to determine when an outbreak or chemical contamination event is occurring solely based on public health surveillance data and other information. Detection necessitates routine monitoring of water quality constituents and contaminants as well as a systematic process for evaluating water quality data sets to determine if an anomaly is occurring. This chapter covers analytical methods used to detect contaminants as well as a discussion of the various components of a contaminant warning system, including online contaminant monitoring and analysis of the resulting data sets using sensor network computational techniques.

2. DETECTION METHODS

Current water quality monitoring strategies were developed to meet regulatory drinking water standards and were not specifically developed to ensure water security and protection against deliberate and accidental water contamination events. Existing monitoring schedules are based primarily on contaminants that are expected to be present in source waters. Standard testing protocols do not include the wide array of biological, chemical, and radiological contaminants that could be used in an attack. In addition, analytical methods necessary to detect the diverse and complex contaminants of concern may not be rapid enough to track outbreaks as they occur. Real-time monitoring is needed to gather immediate feedback.

Technologies to detect contaminants are emerging and being refined (Appendix E). In particular, advancement is occurring in rapid, field testing methods. Ho et al. (2005) provide a comprehensive review of sensors that can be used for environmental samples, including technologies that can be used for detecting trace metals, radioisotopes, and other contaminants. Testing for constituents in drinking water can be challenging, since the contaminants may be present in low concentrations that still have health effects, necessitating methods with low detection limits be adopted. Continued progress in the detection of contaminants is expected.

2.1. Direct Analysis of Contaminants

Detection of specific chemicals that could be used in intentional contamination generally requires use of advanced analytical methods, such as gas chromatography/mass spectrometry, inductively coupled plasma techniques, and ion chromatography (Hall et al., 2007). Some examples of potential chemical contaminants and their corresponding analytical methods are shown in Table 5.1.

Based on the recommendations of Magnuson et al. (2005), the following analytical methods could be useful for detecting chemical and biological contaminants:

- Gas chromatography (GC)
- GC/mass spectrometry (GC/MS)
- Liquid chromatography (LC)
- LC/MS
- Immunoassay test kits
- Ion chromatography
- Graphic furnace atomic absorption
- Cold vapor atomic absorption
- Inductively coupled plasma (ICP)

TABLE 5.1 Analytical Methods for Chemical Contaminants (Magnuson et al., 2005)

Contaminant Category	Specific Examples	Analytical Methods
Cyanide		Colorimetric or ion-selective electrode
Pesticides		Immunoassays
Schedule 1 chemical weapons	VX, sarin	Enzymatic or colorimetric
Volatile organic chemicals	BTEX	Gas chromatography/mass spectrometry (GC/MS), liquid chromatography/mass spectrometry (LC/MS)
Inorganic chemicals	Mercury, lead	Ion chromatography, graphic furnace atomic adsorption, cold vapor atomic absorption, inductively coupled plasma (ICP), ICP/MS
Biotoxins	Ricin, botulinum	Immunoassay
Pathogens	Tularemia, anthrax, plague	Molecular methods (PCR based)

- ICP/MS
- Ion-selective electrodes
- Culture-based microbiological tests
- Molecular-based microbiological tests
- Biochemical and serological tests
- Immunomagnetic separation
- Immunofluorescence assay microscopy

In the event that contamination occurs and the contamination agent is completely unknown, combinations of these analytical methods are needed to try to determine the contamination agent. However, these methods are expensive and require a high level of expertise that is prohibitive for application to routine monitoring.

2.1.1. Biological Agents

For samples containing unknown microbiological contaminants, concentrated samples should be sent to a qualified laboratory for identification (U.S. EPA, 2003b). Both culture-based and molecular-based testing should be done on samples containing unknown biological agents. Culture-based testing involves attempting to grow biological contaminants on selective media; identification can be based on the ability of a culture to grow on specific media and on identification using microscopy. Another screening tool for biological agents is immunomagnetic separation and immunofluorescence assay microscopy. Exploratory methods include those that are molecular-based and rely on polymerase chain reaction (PCR) and probe hybridization. Molecular testing can be used to generate preliminary results that are confirmed using sequence analysis. Molecular-based PCR methods can yield fast results compared to culture-based tests.

Due to the potentially low concentration of pathogenic microorganism in water samples, concentration of field samples using ultrafiltration may be necessary (U.S. EPA, 2003b). Several studies have shown that filtration to concentrate samples is effective for biological assays and field concentration of samples prior to analyses can lower the detection limits to levels that are suitable for drinking water (States et al., 2006). Lindquist et al. (2007) used ultrafiltration with 100 liter samples to detect biological contaminants.

Screening for biotoxins can be done using immunoassays with confirmation by GC/MS, LC, or LC/MS (U.S. EPA, 2003b). A large number of test kits are available, although not all of the kits use standardized methods. These test kits are useful mostly for initial screening. In some cases, animal assays are used for biotoxins and other contaminants that are difficult to detect.

In water and wastewater systems, microbiological safety, as defined by a lack of pathogenic microorganisms, is ensured using indicator organisms, such as total and fecal coliforms or *E. coli*. The presence of indicator organisms typically indicates that fecal material is present, suggesting that pathogenic microorganisms may also be present. In the case of an intentional contamination event, however, pathogens may be used as the contaminant and elevated concentrations of indicator organisms would not be expected.

2.1.2. Chemical Agents

For water samples containing unknown chemical constituents, a large number (e.g., 35) of high-priority contaminants should be tested during screening. Examples of chemical contaminants that should be tested for are included in Module 4 of the U.S. EPA Response Protocol Toolbox (U.S. EPA, 2003b). In general, the following analytical methods can be used to screen for the following chemicals (U.S. EPA, 2003b):

- Volatiles (organic): purge and trap GC/photoionization detector (PID)/electrolytic conductivity detector (ELCD) or GC/MS
- Semivolatiles (organic, includes many pesticides): solid phase extraction GC/MS
- Carbamate pesticides (organic): high-performance liquid chromotography (HPLC), fluorescence detection
- Quaternary nitrogen compounds (organic): HPLC ultraviolet (UV) detector
- Trace metals (inorganic): inductively coupled plasma atomic emission spectrometry (ICP AES), ICP/MS, graphite furnace atomic absorption (AA)

- Total mercury: cold vapor AA
- Cyanides: wet chemistry
- Radionuclides: gross alpha, gross beta, gross gamma

Broader screening for organic chemicals necessitates an array of sample preparation techniques: micro liquid-liquid extraction, continuous liquid-liquid extraction, solid-phase extraction, solid-phase microextraction, headspace collection, and flow injection (U.S. EPA, 2003b). Appropriate analytical methods for organic chemicals could include: multidetector GC in screening mode, gas chromatography with electron impact ionization mass spectrometry, high-performance liquid chromatography UV detector (HPLC UV), high-performance liquid chromatography mass spectrometry (HPLC/MS), tandem mass spectrometry (MS/MS), high-resolution mass spectrometry (HR/MS), and immunoassays (U.S. EPA, 2003b). A more comprehensive screening for inorganic chemicals could include the following analytical methods following sample extraction: ICP AES or ICP MS in semi-quantitative mode, ion chromatography (IC), wet chemistry, and ion selective electrodes (ISE) (U.S. EPA, 2003b). Screening for cyanides necessitates distillation. Improved methods for detecting semi-volatile chemicals have been developed to specifically address chemicals that could be used in an attack on a water system (Grimmett and Munch, 2013).

2.1.3. Radiological Agents

Expanded screening of radionuclides can be guided using the Multi Agency Radiological Laboratory Analytical Protocols manual and the U.S. EPA's Radiological Emergency Response Plan. Radioactive isotopes can be detected using existing methods. Radiation detectors rely on high-purity germanium, scintillation crystals, and Geiger counters (Ho et al., 2005). However, Porco (2010) reviewed currently available detection technologies for radioactive constituents that have the detection limits applicable for monitoring intentionally injected contaminants in drinking water and found that relevant technology is not available.

2.2. Toxicity Testing

Toxicity testing uses biological organisms to detect toxicity in water samples rather than testing for individual constituents. The main advantage of toxicity testing is that it is not necessary to test for the multitude of contaminants that could be present. The basis for using toxicity testing is that, when specific biological species are exposed to toxic constituents in a water sample, they have a measureable biological response. To measure the effect of potentially contaminated water on the organisms, a metric must be used to measure the biological response; examples of metrics include luminescence of algae, metabolic functions, and respiration rates. Toxicity-based biosensors may use water fleas, mollusks, algae, or fish as the biological organisms (States et al., 2003, 2004). Fish ventilatory monitoring is a form of toxicity testing in which the fish response to toxins are observed due to changes in ventilation rate, cough rate, ventilation depth, and movement (van der Schalie et al., 2001, 2004).

While a variety of toxicity tests have been developed, cell-based toxicity tests show more favorable results (Curtis et al., 2009a, 2009b; Davila et al., 2011; Eltzov and Marks, 2010, 2011; Giaever and Keese, 1993; Iuga et al., 2009). Eltzov and Marks (2011) provided a comprehensive review of whole-cell biosensors that are appropriate for a variety of applications, including water security. Bacterial cell toxicity tests have compared favorably with other types of toxicity tests that use higher-order biological indicators, although results from these tests do not exhibit a strong relationship with human and animal health data (Botsford, 2002).

While toxicity testing is typically done in a laboratory setting, online biosensors are now available. Comparisons of the commercially available biosensors Eclox™, Microtox™, and ToxAlert™ were made in addition to a toxicity test using *Daphnia magna* by Dewhurst et al. (2002), who found that some of the results were sensitive to changes in water quality and temperature. The Microtox test uses *Vibrio fischeri* as the biological indicator. Some biosensors can be used for rapid on-site testing (States et al., 2004). Biosensors have been shown to be reliable and to generate reproducible results (Campbell et al., 2007; Hillaker and Botsford, 2004; States et al., 2004). A disadvantage of biosensors is that they may not be suitable for testing drinking water because the organisms used in these tests may be sensitive to disinfectants and to other drinking water constituents (Hall et al., 2007).

Van der Schalie et al. (2006) investigated ten toxicity sensors: an electric cell-substrate impedance sensor, Eclox acute toxicity sensor (Severn Trent Services), hepatocyte low-density lipoprotein uptake, Microtox (Strategic Diagnostics, Inc.), Mitoscan (Harvard Bioscience, Inc.), neuronal microelectrode array, *Sinorhizobium meliloti* toxicity test, SOS cytosensor, Toxi-Chromotest (Envrionmental Biodetection Products Inc.), and ToxScreen II (Checklight Ltd.). The tests involved using 13 industrial chemicals to challenge the efficiency of the toxicity monitors: aldicarb, ammonia, copper sulfate, mercuric chloride, methamidophos, nicotine, paraquat dichloride, phenol, sodium arsenite, sodium cyanide, sodium hypochlorite, sodium pentachlorophenate, and toluene. The study by van der Schalie et al. (2006) revealed that none of the sensors responded to nicotine, no single sensor responded to more than six of the chemicals, and that a combination of sensors responded more favorably than using a single sensor.

An example of toxicity testing is the Bioscan™ online sensor, which detects contamination by measuring the biological response of a bacterial culture that is exposed to the water (Campbell et al., 2007). Curtis (2009b) introduced a portable benchtop mammalian cell-based toxicity sensor that uses electric cell-substrate impedance sensing technology. Cells are grown on the fluidic biochips, which use an automated media delivery system to maintain cells for at least nine days (Curtis et al., 2009b). Silicon-based microbial fuel cells can be used as portable biosensors; the variation in the current produced by cells changes when toxicants are present (Davila et al., 2011). Another example of a biosensor is presented by Dierksen (2004), in which a chromatophore-based cytosensor was developed using cells from fish and relying on changes in optical density for toxicity detection.

Toxicity testing can be a useful test method although the results can be inconsistent, depending on the application. The results of toxicity test using different biological indicators can vary depending on the organism used (Radix et al., 2000). Zurita et al. (2007) performed toxicity testing for a potent pesticide and found that the results varied according to which indicator was used. Selection of an appropriate cell line is an important design consideration for making the sensors more sensitive (Curtis et al., 2009a). In one study, the Microtox test compared favorably with other toxicity tests that used rats, ducks, quail, sand fleas, and trout fingerlings for water contaminated with herbicides (Hillaker and Botsford, 2004). However, in another study, a rapid toxicity test using luminescent bacteria was more sensitive than the Microtox test (Ulitzur et al., 2002).

Online toxicity testing has proven to be successful. Using a combination of analytical methods (HPLC-UV and GC-MS) and online toxicity monitoring, de Hoogh et al. (2006) detected an introduced chemical contaminant in a surface water source in the Netherlands. The online toxicity monitor used was the bbe *Daphnia* toximeter, which measured behavioral changes in the test organism based on swimming behavior (de Hoogh et al., 2006). The source of the contaminant was not determined.

Despite the promise of online toxicity testing, the technology is still in development. There have been several attempts to make the technology more accessible and ready for field applications. For example, Jeon et al. (2008) incorporated automation into toxicity testing using *Daphnia magna*. Whole cell fiber optic–based biosensors consist of whole cells that are attached to optic fibers, where a variety of techniques are used to detect biological responses to toxicants (Eltzov and Marks, 2010). However, in their review of fiber optic–based cell-based biosensors, Eltzov and Marks (2010) concluded that a lack of commercial applications is limiting the development of biosensors.

2.3. Novel Detection Technology

In a review following a meeting of experts in the field of water security, Foran and Brosnan (2000) identified the following new technologies for detecting biological contaminants in real-time:

- DNA microchip arrays
- Immunologic techniques

- Microrobots
- Optical techniques
- Flow cytometry
- Molecular probes

In a review of novel and established sensor technology, Ho et al. (2005) described many sensors that are available and being developed to detect a variety of environmental contaminants. For monitoring trace metals, emerging technologies include nanoelectrode arrays, laser-induced breakdown spectroscopy, and miniature chemical flow probe sensors (Ho et al., 2005). For monitoring radioisotopes, the following technologies are being developed: RadFET (radiation field-effect transistor), cadmium zinc telluride detectors, low-energy pin diodes beta spectrometers, thermoluminescent dosimeters, isotope identification gamma detectors, and neutron generators for nuclear material detection (Ho et al., 2005). Monitoring volatile organic compounds (VOCs) has been evolving to include the following technologies: evanescent fiber-optic chemical sensors, grating light reflection spectroelectrochemistry, miniature chemical flow probe sensors, SAW chemical sensor arrays, MicroChemLab (gas phase), gold nanoparticle chemiresistors, electrical impedance of tethered lipid bilayers on planar electrodes, MicroHound, Hyperspectral imaging, and chemiresistor arrays (Ho et al., 2005). Biological sensors being developed include fatty acid methyl esters analyzers, iDEP (insulator-based dielectrophoresis), Bio-SAW sensors, μProLab, and MicroChemLab (liquid) (Ho et al., 2005).

The lab-on-a-chip concept, consisting of pocket-sized chemistry equipment used in combination with portable electronic equipment, is very appealing for a variety of applications, including drinking water security (Gardeniers and Van den Berg, 2004; Jang et al., 2011). The lab-on-a-chip devices could easily be used for field screening, with subsequent testing occurring in the laboratory to confirm the initial results.

Advancements are being made in the detection of biological contaminants, although many of the methods have focused on detection of *E. coli*. Bukhari et al. (2007) investigated rapid test methods for *E. coli*, which included four immunological tests and a molecular assay. Geng et al. (2011) used molecular-based electrochemical biosensors for detecting *E. coli*, a method where magnetic beads are coated to function as DNA probes. Serra et al. (2005) tested an enzyme-based biosensor for one-day detection of *E. coli* and found that the detection limit of 10^6 cfu/ml was improved by adding a preprocessing step for samples to improve detection limits to 10 cfu/ml. Ercole et al. (2002) also investigated a biosensor for *E. coli* based on an immunoassay test, where detection is based on detecting pH variations due to ammonia production by an antibody reaction.

Additional methods have been used to detect biological contaminants. Xie et al. (2005) used a combination of laser tweezers and confocal Raman spectroscopy to detect single bacterial cells in aqueous solutions using a rapid method. Sengupta et al. (2006) used surface-enhanced Raman spectroscopy with a suspension of nanocolloidal silver particles to detect *E. coli* with a detection limit of 1000 cfu/ml. Shoji et al. (2000) investigated a biosensor that measures the uptake of fluorescently labeled proteins by cultured human cells.

In addition to advancements in biological contaminant testing, advances have been made to better detect trace metals and other chemical constituents. Many of these tests and devices are being developed to address needs other than security (e.g., water quality issues such as arsenic in well water); however, these tests may also serve in enhanced water security. Examples include microfluidic sensors that can detect low concentrations of contaminants such as lead (Chang et al., 2005), a biosensor used to detect arsenic in drinking water (de Mora et al., 2011), micro-chemiluminescence enzyme tests for mercury in drinking water (Deshpande et al., 2010), a self-powered enzyme biofuel cell on a microchip for cyanide detection (Deng et al., 2010), and a biosensor for arsenic detection that is inexpensive and field deployable (Joshi et al., 2009). Advancements have also been made in testing for lead in drinking water (Wang et al., 2001).

Additional devices are being developed to test for various contaminants of concern in drinking water systems. Yang et al. (2010) present a nanotube array method for detecting hydrocarbons in surface water. A polymer-coated surface-acoustic-wave microsensor was developed as a cost-effective way to detect VOCs in drinking water on-site (Groves et al., 2006). A handheld device for detecting gas-phase contaminants, MicroChemLab, was developed by Sandia National Laboratory (Lewis et al., 2006). The MicroChemLab device consists of a micromachined preconcentrator, a GC channel, and a quartz surface acoustic wave array detector (Lewis et al., 2006). Other technologies for detection are currently under development.

2.4. Standards for Analytical Methods

Many standards are available for contaminants typically found in drinking water and environmental samples; however, testing for intentionally added contaminants may require the use of hazardous materials testing protocols and unconventional analytical methods. Nonetheless, standard methods are still useful for detecting certain contaminants. A resource for standard analytical methods is the database titled the National Environmental Methods Index for Chemical, Biological, and Radiological Methods (NEMI-CBR). Additional methods are published by U.S. EPA, American Public Health Association, American Society for Testing and Materials, Association of Analytical Communities, and International Organization for Standardization. The U.S. EPA Response Protocol Toolbox contains a list of standardized methods that can be useful in preparing for and responding to water threats (Table 4-2 in the RPTB, U.S. EPA, 2003b). Federal laboratory networks and various state and federal agencies provide information on proper sample collection protocols and concentration methods

(Crisologo, 2008; States et al., 2006; U.S. EPA, 2003a, 2003b). Chapter 6 contains more information on laboratory resources and networks.

2.5. Rapid On-Site Test Kits

Magnuson et al. (2005) provided a recommended battery of tests to be performed during a suspected event to verify the presence of contaminants. While this strategy does not represent real-time detection, it is an important component of a CWS and could be used to complement real-time detection efforts. Magnuson et al. (2005) recommend two types of field kits for detecting contamination events (Table 5.2). The regular field kit represents tests and equipment that could easily be obtained and used by water providers. The expanded field kit contains more specialized tests that require more investment and training on the behalf of the water provider using the field kit. When using field kits to test water for intentional contamination, it is imperative to practice safe emergency response protocols with highly trained individuals. Safety of employees deployed into the field to test water parameters is paramount and may slow response activities. Guidelines for basic and expanded screening for contaminants is contained in the U.S. EPA Response Protocol Toolbox (Figure 4-6, U.S. EPA, 2003b).

The collection of samples for analysis should be done in accordance with standard methods. Concentration of samples may be necessary prior to analysis to detect target microorganisms with low detection limits (Magnuson et al., 2005). As part of the Environmental Technology Verification (ETV) Program, eight sensor technologies with the potential to detect the presence of contaminants in drinking water were examined:

- IQ Toxicity Test™ (Aqua Survey, Inc.)
- Deltatox® and Microtox® (Strategic Diagnostics Inc.)
- Eclox (Severn Trent Services)
- PolyTox™ (InterLab Supply, Ltd.)

TABLE 5.2 Parameters to Test When Evaluating a Threat (Magnuson et al., 2005)

Recommended Field Kit	Recommended "Expanded" Field Kit
Radioactivity: Geiger-Muller probe	General hazards (performed by trained
Cyanide: colorimetric test or probe	hazardous materials responder)
Chlorine residual: colorimetric test	Volatile chemicals
pH: probe	Schedule 1 chemical weapons
Conductivity: probe	Water quality parameters
	Volatile organic compounds
	Biotoxins
	Pathogens
	Toxicity

- ToxTrak™ (Hach)
- ToxScreen (CheckLight, Ltd.)
- BioTox™ (Hidex Oy)

Additional testing of rapid on-site tests has been completed by other groups. Several of the preceding tests were also evaluated by van der Schalie (2006). States et al. (2004) extensively tested the following rapid field tests:

- Bio-Threat Alert® (Tetracore) test strips that are based on lateral flow immune chromatography.
- SMART (Sensitive Membrane Antigen Rapid Test) (New Horizons Diagnostic) tickets that detect antigens by immune focusing colloidal gold-labeled reagent (antibodies) and target antigens onto membranes.
- Rapid enzyme test (Severn Trent Services) test strip consisting of a membrane disk saturated in cholinesterase.
- R.A.P.I.D. (Ruggedized Advanced Pathogen Identification Device) (Idaho Technology) field-deployable PCR system.
- HAPSITE GC/MS (INFICON Inc.) field-deployable system.
- Deltatox® and Microtox® (Strategic Diagnostic Inc.) toxicity tests, Deltatox® is the portable field version and Microtox® is the laboratory version.
- Eclox (Severn Trent Services) toxicity test.

States et al. (2004) found that the rapid test kits were useful for detecting contaminants but that all the tests had some limitations and none was a substitute for laboratory analysis. One of the challenges of using these rapid tests is that false-positive and false-negative results can occur, which make interpretation of the results more complicated (Hrudey and Rizak, 2004).

2.6. Water Quality Parameters for Detection of Contaminants

All possible biological, chemical, and radiological threats represent a large group of constituents. The number of possible contaminants is so large that it is not feasible to simultaneously monitor all these constituents in water systems. In addition, the contaminants of concern are constantly changing. Although it is not possible to monitor in real-time all or even a large percentage of the total contaminants that could be present, practical strategies for dealing with the inability to measure all of the contaminants of concern include measuring only the most important or likely contaminants. An alternative approach is to measure commonly monitored water quality parameters, such as pH, temperature, and electrical conductance, and anticipate that these parameters will register as abnormal values if there is contamination. The following are standard water quality parameters and operating conditions that could represent potential contamination if values deviate greatly from baselines:

- Alkalinity
- Carbon: total inorganic (TIC) and organic (TOC)

- Chloride
- Chlorine residual: free and total
- Color
- Conductivity
- Dissolved oxygen (DO)
- Fluorescence
- Nitrogen: ammonia, nitrate, nitrite
- Odor
- Oxygen-reduction potential (ORP)
- Particle size distribution
- pH
- Pressure changes and abnormalities
- Taste
- Temperature
- Total dissolved solids
- Turbidity
- UV absorbance (e.g., UV254).

Monitoring standard water quality parameters and operating conditions have many advantages. Monitoring technology for these constituents is readily available and relatively inexpensive compared to technologies for monitoring specific contaminants. Many of these constituents can be measured in real time. Specialized facilities are not necessary and employee training is minimal. Due to the low cost of these technologies, it may be possible to deploy more of the devices throughout the water system and collect data in real time. Operations data management systems are already used by operations personnel to track operational data that could be used to establish the baseline operating conditions necessary for detection of contamination and calibration of hydraulics models of distribution and treatment systems (Serjeantson et al., 2011). Smeti et al. (2009) used a statistical approach to predict water origin in a water system, suggesting that water quality data can be used to predict origin as it refers to a contamination location.

Of the water quality parameters tested, chlorine residual has been shown to be an effective indicator of several contaminants. In laboratory tests, Helbling and VanBriesen (2007) demonstrated that microbial suspensions, a possible method of attack, exert a chlorine demand and that changes in chlorine residual could be evidence of contamination.

3. CONTAMINANT WARNING SYSTEMS

One of the most important elements in developing a water security program is to design and implement a contamination warning system (CWS) to detect contamination events so that an appropriate response can follow. The U.S. EPA worked on the development of a CWS as part of the Water Security Initiative that was initially referred to as the WaterSentinal project (U.S. EPA, 2007,

2008b, 2008c). In addition, Murray et al. (2008) summarized the framework for designing a CWS, including guidance on determining the number of sensors used and determining system objectives. A comprehensive CWS should consist of the following elements (U.S. EPA, 2007):

- Online contaminant and water quality monitoring.
- Sampling and analysis.
- Enhanced security monitoring.
- Consumer complaint surveillance.
- Public health surveillance.

Each of the five elements of a CWS involves a concentrated effort by water utilities. Deployment of online water quality monitoring should include online monitoring sensors located throughout the distribution system and collection of water quality data in real time. Data from the sensors should automatically be electronically transferred to a centralized network for storage and analysis. Software is needed to analyze and interpret the sensor data sets. Sampling protocols should include routine sampling and additional sampling when a threat is possible. Enhanced security monitoring consists of equipment and devices that are intended to physically protect critical assets. Examples include video cameras, motion triggered lighting, alarms, motion detectors, and hatch intrusion alarms. Consumer complaint data can be useful for determining if a contaminant is present if the odor, taste, or color of drinking water is affected. Public health data can also be useful and examples of data that can be collected include: OTC drug sales, hospital admissions reports, and 911 calls.

An effective CWS can result in multiple benefits. A well-designed CWS can also assist in locating the source of contamination during a contamination event and assist in carrying out recovery and rehabilitation efforts to restore contaminated distribution systems. Installation of a CWS in an operational water system makes real-time response possible. Additional benefits of a CWS include better capabilities for understanding the effect of operational changes on water quality and better dosing of disinfectant residuals throughout the distribution system. In addition to providing information about water quality events, CWSs can be used to detect other types of threats. For example, instead of investigating contamination events, Skolicki et al. (2006) considered the effects of breaks in pipelines using a sensor network approach.

3.1. Online Contaminant Monitoring

Real-time detection of contamination can be achieved using online monitoring stations. Researchers demonstrate the benefits of online monitoring over routine sampling and analysis (Janke et al., 2006). In developing a CWS, the following needs to be determined: number of monitoring stations, parameters to measure, analytical methods, locations of monitoring stations, water distribution modeling software to use, and data analysis methods (Rober-son and Morley, 2005). The water distribution modeling software must be dynamic to simulate flow and contaminant transport throughout the distribution system. In contrast, water utilities typically use static models for planning purposes.

Prior to deploying a CWS, water utilities need to develop standard operating procedures (SOPs) to determine appropriate responses that should follow each alarm and trigger (U.S. EPA, 2007). First, sufficient data must be collected following installation of online monitoring stations to establish baseline conditions. At least one year of data should be collected and analyzed to determine seasonal fluctuations. Next, triggers that indicate deviations from the baseline and abnormal conditions must also be established. A trigger indicates a possible threat that must be evaluated to determine if it is credible. When evaluating anomalies, the following need to be considered: absolute magnitude of changes, magnitude of changes relative to the size and fluctuations in the baseline, and the slope of the change. Ideally, contaminant warning systems should include monitoring that is "sensitive, specific, reproducible, and verifiable" (Foran and Brosnan, 2000 p. 993). All data generated by monitoring stations needs to be compiled and evaluated using a systematic QA/QC program. Although much of the data analysis will likely be completed automatically, manual input regarding alarms and triggers will need to be entered. Rigorous review of the data is needed before a threat is determined to be possible, since false-positive and false-negative errors are problematic with online contaminant monitoring technology (Magnuson et al., 2005).

In addition to the ability to detect contamination events, considerations for CWS should include cost, ease of use, automation, and reliability (Kroll and King, 2010). An ideal online contaminant monitoring system has the following characteristics (Roberson and Morley, 2005):

- Rapid response time
- Automated
- Ability to preserve samples
- High sampling frequency
- High sensitivity for public health threats
- High specificity for a broad spectrum of contaminants
- High reproducibility
- Low rate of false positives and false negatives
- Rugged construction
- Ease of use
- Inexpensive design
- Low operation and maintenance requirements
- Low power demand

The characteristics of an ideal CWS are challenging to implement (U.S. EPA, 2007). No single water constituent can be used to truly detect a broad spectrum of contaminants. Operating a CWS with a minimal number of monitoring stations suggests that each monitoring station needs to be able to cover an expansive portion of the distribution system. Due to the time-sensitive nature of detecting contamination and formulating a response, an online monitoring system must detect constituents quickly and facilitate a timely response. To gain support from water utility staff members and customers, the online contaminant monitoring system needs to produce a minimal number of false-positive errors. Finally, the system needs to be a sustainable long-term solution with benefits in addition to providing better security.

Research and development of a CWS is an ongoing task, with development of new sensors and strategies to analyze sensor data sets underway. Significant effort has been made to simulate potential contamination events and field test CWSs in water distribution systems. The U.S. EPA tested CWSs in simulated drinking water distribution systems. The U.S. EPA has also sponsored pilot demonstration projects for CWS in three cities. The first pilot demonstration was completed in Cincinnati, Ohio, and was well-documented (Allgeier et al., 2011; Fencil and Hartman, 2009; Pickard et al., 2011; U.S. EPA, 2008a).

To advance development of CWSs, the U.S. EPA developed the Threat Ensemble Vulnerability Assessment (TEVA) program, where volunteer water utilities used utility-specific dynamic water quality distribution system models with contaminant information to assess the potential consequences of contamination and potential mitigation measures. The TEVA research team consisted of the U.S. EPA, Sandia National Laboratory, University of Cincinnati, and Argonne National Laboratory. Murray et al. (2009) estimated that the CWS developed for utilities participating in the TEVA program could reduce fatalities expected during a contamination event by 48% and reduce associated economic consequences by over $19 billion.

A CWSs has beneficial uses in addition to detecting contaminants. Hart and Murray, (2010) suggested the following ways that sensors and sensor networks can be used for purposes other than detecting intentional contamination events:

- Maintaining chlorine residual throughout the service area.
- Performing routine required sampling.
- Monitoring disinfection by-products.
- Monitoring pressure.
- Detecting leaks.

3.2. Online Sensor Technology

The adaption of water quality sensor devices and integration of these sensors in a water distribution system is limited by the technologies available and financial resources. No currently available sensors are capable of directly detecting a large number of potential contaminants and are economical. Sensors that monitor water quality parameters and are deployed in a water distribution system may be useful for detecting contamination events (Hall et al., 2007). In addition, sensors need to be developed that are both economical and reliable to expand the capabilities to detect contamination events (States et al., 2003).

A variety of sensor technologies are used for online contaminant monitoring stations. Some of these technologies are based on standard water quality parameters and measure either singular parameters or a host of parameters. In addition to readily available and well-documented technologies, sensors based on emerging technologies are being developed, although they are not yet market-ready. The sensors use tests for water quality parameters, toxicity, and properties indicating microbial presence (e.g., particle counts). Biosensors are available that use fish, algae, and bacteria. Note that deployment of these biosensors involves considerable effort on the part of the user to provide the biological species necessary to indicate toxicity. Biosensors are nonspecific and target toxicity or microorganisms rather than attempting to target specific contaminants. Although some of the microbiological sensors have been developed to detect *E. coli* and other common bacteria, it is possible that these sensors could be modified to detect additional microorganisms in the future.

Although online multivariate probes are useful for tracking water quality in distribution systems, unsteady hydraulic conditions can affect results and sensor accuracy can be unreliable (Aisopou et al., 2012). In particular, Aisopou et al. (2012) found that resuspension of sediments and mixing conditions in pipes was problematic for detecting disinfectant residual, turbidity, and color. Sources of variation in water quality data include instrument noise, change in instrument response over time, actual water quality variation (seasonal, residence time), and distribution system operational changes (source water blending, valve opening, pipe flushing) (Murray et al., 2012).

In a review of online contaminant monitoring technology, Storey et al. (2011) found no universal detectors available and most novel sensors are not ready for full-scale implementation. Storey et al. (2011) reviewed the following technologies: JMAR Biosentry™, UV-VIS s::can spectro::lyser™, Hach Event Monitor (Guardian Blue™), YSI Sonde™, Censar™, s::can Water Quality Monitoring Station™, TOXcontrol™ (microLAN), Algae Toximeter (BBE), *Daphnia* Toximeter™ (bbe), ToxProtect™ (bbe), Fish Activity Monitoring System (FAMS), surface enhanced Raman spectroscopy, laser tweezer Raman spectroscopy, and surface acoustic wave devices. Cost is one of the biggest issues in selecting online sensors. In addition, dual-use benefits are required for water utilities to justify the investment.

3.3. Sensor Network Hardware

Implementation of a CWS involves significant information technology (IT) components that need to be integrated into the existing IT system, including the transmission, management, and storage of electronic data sets (U.S. EPA, 2007). Many water utilities currently operate using SCADA systems to remotely monitor and control facilities, provide transmission of alarms and notifications, and provide automation capabilities and safety checks. The data from CWS sensor monitoring stations should be integrated into the SCADA system to provide a streamlined organizational structure and buy-in from water utilities staff. One challenge associated with implementing a CWS is developing the capability to transmit and analyze large data sets. Options for data transmission include hardwired systems (wire, coxial cable, fiber optics) or wireless systems (microwave, radio, cellular, WiFi) (Roberson and Morley, 2005).

3.4. Sensor Network Software

Once online contaminant and water quality monitors are deployed in a water distribution system, methods need to be developed to manage and analyze the resulting data sets, which are typically large, since the sensors are continuously collecting data. A sensor network approach is often taken to manage the large data sets generated. Many researchers have investigated the use of optimization protocols for the design of sensor networks and for contamination detection. Sensor network design typically involves determining the number of sensors to place in the network and the optimal locations for these sensors. Review papers summarizing the use of optimization protocols and development of CWS are available (Ailamaki et al., 2003; Murray et al., 2008, 2009).

Adapting and applying sensor network software to CWSs for water distribution systems can have multiple benefits. The software can aid in sensor network design. The software can be used to detect contamination as well as water quality occurring as the result of operational conditions. Use of the software can also include simulating contamination events to aid in planning and vulnerability assessments. During a water threat or confirmed contamination, the software can be helpful in responding to and carrying out rehabilitation efforts. When modeling water systems using computer software, some utilities model all pipes in the systems, while others model only larger pipes and critical infrastructure (Hart and Murray, 2010).

Running simulations of water contamination events can be used to better understand threats, although a variety of contaminants and scenarios need to be considered. Janke et al. (2010) found that highly toxic contaminants caused effects that were sensitive to injection duration, while less toxic contaminants caused effects that were more strongly influenced by total injection mass. In simulating contamination events, Khanal et al. (2006) found that water demand and injection mass influence the impact on the population the most, while the injection duration was less important.

Most published papers on sensor networks represent work that was performed on hypothetical water systems and not on real water systems (Hart and Murray, 2010). However, more studies are being conducted using data from real water systems. Alfonso et al. (2010) tested use of EPANET with an optimization formulation to determine how to conduct flushing in contaminated water systems; the approach was tested using data from the full-scale water system in Villavicencio, Colombia.

Software packages are available to apply sensor network approaches to water distribution systems. The EPANET software was developed by the U.S. EPA and can be used to study contamination in water distribution systems. This software, which is free and available online, can model hydraulic and water quality conditions in pressurized pipe networks. Users can enter information about their water system and study transport of reactive chemicals using an EPANET software add-on called EPANET-MSX. An additional add-on called AZRED allows users to model hydraulics and contaminant transport in distributions networks where incomplete mixing occurs at pipe intersections. Another extension of EPANET is PipelineNET, which combines water distribution system hydraulic and water quality modeling with geographical information systems (GIS) by integrating EPANET using ArcView GIS software.

While EPANET is used to model hydraulic and contaminant conditions in pressurized water distribution systems, other computer models are available for modeling contamination in the source water. An example is RiverSpill, which is a GIS-based contaminant transport model (Samuels et al., 2006). Samuels and Bahadur (2006) advocate an integrated multiple contaminant tracking approach for potential contamination in source waters and in distribution systems using RiverSpill, PipelineNET, and ICWater (Incident Command Tool for Drinking Water Protection). An advantage of using GIS-based modeling methods is that water utilities may already be using these methods for oversight of their distribution system.

3.4.1. Optimal Sensor Placement

One of the most important decisions in designing a CWS is to determine how many sensors are needed and where they should be located. Sensors should be placed where they will be most effective for detecting contamination events. Designing a system with the least number of sensors possible allows for a system that is effective as well as economical. Designing a CWS is not as simple as performing a single sensor placement analysis. Practical constraints, capital, operation and maintenance costs, and the priorities of the water utility and municipality may make it difficult to identify a single best sensor network design. Consequently, the design process requires informed decision making, where sensor placement methods are used to identify possible network designs that work well under different assumptions and for different objectives.

In a review of over 90 studies related to optimization of sensor placement in water networks, Hart and Murray (2010) summarized the following methods for determining the optimal placement of sensors:

- Human judgment based on expert opinion to estimate locations where water quality is representative of entire system.

- Ranking methods that offer a more methodical approach to estimating where sensors would be most optimally placed.
- Optimization using modeling, which is the most defensible approach.

Sites for online contaminant monitoring stations need to have access to water, drains, power, and communication systems (U.S. EPA, 2007). In addition, these sites should be secure and accessible. Locating sensors near existing larger infrastructure such as reservoirs, pump stations, and wells can be beneficial, since these sites represent critical points within the distribution system and already have power and communication infrastructure. It may also be advantageous to locate sensors near critical facilities, such as hospitals (Roberson and Morley, 2005).

Optimization models are conducive to the design of sensor networks because multiple objectives can be met simultaneously. In the context of water security and CWSs, the following network design objectives might be used (U.S. EPA, 2007):

- Minimize health consequences to the population.
- Minimize the spatial extent of contamination.
- Minimize the time until detection occurs.
- Maximize the spatial coverage of sensors throughout the distribution system.
- Maximize the number of contamination events detected in simulations.
- Maximize protection of key facilities and populations.

Threat Ensemble Vulnerability Assessment Sensor Placement Optimization Tool

To facilitate optimal sensor placement within a network, the Threat Ensemble Vulnerability Assessment Sensor Placement Optimization Tool (TEVA-SPOT) was developed by the U.S. EPA, Sandia National Laboratories, Argonne National Laboratory, and the University of Cincinnati (U.S. EPA, 2012b). The purpose of TEVA-SPOT is to determine the optimum number of sensors and placement of these sensors within a distribution system (Murray et al., 2009). TEVA-SPOT works in conjunction with EPANET. Using TEVA-SPOT, optimization of sensor network design occurs based on the number of people exposed to dangerous levels of a contaminant, volume of contaminated water used by customers, number of feet of contaminated pipe, and time to detection (U.S. EPA, 2010b). An advantage of TEVA-SPOT is that it can be run with multiple objectives. It is based on mixed integer programming (MIP) and can be formulated with imperfect sensors, supporting sensor failure (U.S. EPA, 2012b). By inverting the sensor placement problem, the minimum number of sensors needed to provide a specified level of protection can be determined.

The TEVA-SPOT software was used to design a sensor network for the city of Ann Arbor, Michigan (Skadsen et al., 2008). The city chose to implement four stations for security and four stations for water quality assurance. Only one location housed both types of stations, so a total of seven sites were selected. Although the city evaluated six parameters for security and water quality concerns (chlorine residual, UV254, ammonia, chloride, dissolved oxygen, and conductivity), only chlorine residual, DO, conductivity, and UV254 were used at all stations.

Computational Approaches

Significant progress has been made over the last ten years in developing analytical capabilities to design sensor networks for CWSs (Hart and Murray, 2010). Earlier efforts in computational sensor placement relied on steady state models (those prior to approximately 2005), but more recent efforts include more complex dynamic contaminant transport mechanisms (Hart and Murray, 2010). Many early examples can be found of sensor placement in water distribution systems (Kessler et al., 1998). Influential work in sensor placement was completed by Berry et al. (2005) and Berry et al. (2006). In Berry et al. (2005), EPANET software was used along with ILOG's AMPL/CPLEX mixed integer program to solve the sensor placement optimization problem. Optimization is based on minimizing the number of people exposed and the contaminant injection is assumed to occur at a single point. The work of Berry et al. (2005) was updated by Berry et al. (2006) to include temporal characteristics of contamination events by including an explicit external transport formulation. Contaminant concentrations are calculated over time at specified locations within the network. In the work of Berry et al. (2006), a MIP solver was used in addition to a Greedy Randomized Search Procedure heuristic method for p-median formulation. The added complexity of the temporal sensor problem makes its solution more computationally intensive.

Multiple approaches have been used to solve the inherently underdefined optimization problem of locating sensors in networks where only a limited number of sensors are feasible. There are many reports in the literature of solving these optimization models (Ostfeld and Salomons, 2004, 2005; Propato, 2006). Over time, the optimization problems have evolved from single objective functions to multiobjective functions. Consideration of delays in responding to events is also included in the model formulation by Berry et al. (2009).

Preis and Ostfeld (2008c) apply a multiobjective sensor placement approach that takes sensor detection, redundancy, and response time into account. The likelihood of sensor detection is optimized using the following equation:

$$f_1 = \frac{1}{S} \sum_{r=1}^{S} d_r \qquad (5.1)$$

where S is the total number of contamination scenarios and f_1 is maximized (Preis and Ostfeld, 2008c). The variable $d_r = 1$ for a detected contamination event, r, under specified flow, concentration, and duration conditions, but is 0 if the contamination is not detected (Preis and Ostfeld, 2008c). Redundancy in the sensor network is addressed by the following equation:

$$f_2 = \frac{1}{\sum_{r=1}^{S} d_r} \sum_{r=1}^{S} R_r \tag{5.2}$$

where f_2 is maximized and R_r is the redundancy of the sensor network for the rth contamination scenario, a 1 or 0 value is defined as requiring at least three sensors up to 30 minutes from the first to the third detection of a nonzero contaminant (Preis and Ostfeld, 2008c). The time for any sensor in the system to detect an event, t_d, is given as

$$t_d = \min_i t_i \tag{5.3}$$

where t_i is the time of detection by the ith sensor (Preis and Ostfeld, 2008c). The time of sensor detection is optimized using the equation:

$$f_3 = E\left(t_d\right) \tag{5.4}$$

where f_3 is minimized, and $E(t_d)$ is the mathematical expectation for t_d (Preis and Ostfeld, 2008c).

Some researchers based sensor design on extreme events. Perelman and Ostfeld (2010) evaluated sensor design to extreme impact contamination events and compared this approach with previous work. Watson et al. (2009) used an approach similar to that of Berry et al. (2006), but they consider protection against high-consequence events. The alteration in the optimization problem does increase the complexity and computational requirements. In addition, by designing for high-impact events, the protection for lower-impact events is reduced.

While earlier models assume that the sensor gives perfect results, later efforts incorporate occurrences of false-positive and false-negative results into their formulations. Preis and Ostfeld (2008a) investigated imperfect sensors in their optimization model, including sensors with limited resolution and sensors with binary output. Imperfect sensors are also considered in the work of Koch and McKenna (2011). In the work of Berry et al. (2009), the possibility of imperfect sensors was also included in the model formulation by assigning a probability that false-negative results would be reported by the sensors. The main conclusion of Berry et al. (2009) was that even imperfect sensors are beneficial.

Various researchers use varying approaches to solve the sensor placement problem. Chang et al. (2011) used a rule-based expert system (RBES) in lieu of the typically used optimization formulation to reduce the computational resource requirements of the sensor placement problems. The two rules used are based on accessibility and complexity. Results from the RBES approach compare favorably with results obtained using optimization solutions. Cozzolino et al. (2011) used an impact probability distribution (IPD) with an external-temporal formulation, similar to that of Berry et al. (2006). Eliades and Poly-carpou (2010) presented the mathematical formulation for fault monitoring to locate sensors in networks. Dorini et al. (2010) utilized a sensors local optimal transformation system (SLOTS) that addresses both single objective and multiobjective approaches to optimization based on detection likelihood and population affected prior to detection; this approach appears to outperform the "greedy" algorithm. Krause et al. (2008) used the concept of submodularity to optimize sensor design by using the concept of penalties and solving to minimize penalties. Shen and McBean (2011) used a nondominated genetic algorithm (NSGA-II) to solve the optimization problem with a Pareto optimality methodology. Many other researchers use the Pareto front principle to aid in optimization problem solutions. Xu et al. (2010) used a robust placement formulation with scenario-based minimax and minimax regret models and a heuristic solution algorithm. Afshar and Marino (2012) solved the sensor placement problem using a nondominated archiving ant colony optimization (NA-ACO) approach, which is a multiobjective ant colony–based algorithm.

Some researchers attempt to simplify sensor placement formulations and reduce computational requirements. Rico-Ramirez et al. (2007) used a two stage optimization process with a stochastic decomposition algorithm and considered sensors that have high and low costs. Including cost in the sensor placement problem is beneficial, since there does appear to be a disparity among sensor technologies in terms of complexity and cost. Propato (2006) used data preprocessing to reduce the sensor placement problem and computational effort.

Most research efforts in sensor placement use EPANET software as the basis for investigation; however, other approaches have been used. Weickgenannt et al. (2010) used ALEID simulation software, which is based on EPANET. Isovitsch and VanBriesen (2008) integrated GIS methods into the sensor placement problem and relied on spatial relationships instead of optimization algorithms. The results compare favorably with those from optimization studies, and the problem solution is less computational and uses GIS software, which is more familiar to the water utilities.

Initially, only a few sensor placement techniques were applied to real water systems (Hart and Murray, 2010). However, application of sensor design to real water systems appears to be more common. Weickgenannt et al. (2010) applied the sensor placement problem to the water system located in the city of Almelo, in the Netherlands. One of the conclusions made by Weickgenannt et al. (2010) was that there is a reasonable benefit from a small number of sensors and diminishing returns for a large number of sensors. Shen and McBean (2011) applied the sensor placement problem to the water system located in the city of Guelph, Canada (population 110,000).

Interest in designing sensor systems is reflected in some of the competitions developed to investigate multiple computational techniques. In The Battle of the Water Sensor Networks (BWSN), an event initiated at a conference in 2006, researchers were challenged to develop network designs and test their designs on the same hypothetical systems. Ostfeld et al. (2008) reviewed the results of the BWSN in which 15 independent groups submitted their results for the challenge. The water system conditions for the BWSN were defined in advance. The participants used varying solutions to solve this optimization problem. The solutions presented to the BWSN were variable, highlighting the multiobjective nature of locating sensors. In a red team/blue team exercise, some participants (red team) acted as the attackers and other participants (blue team) acted as the defenders; one conclusion from this exercise was that it is important to minimize delays in detection in order to increase the likelihood of detection (Grayman et al., 2006).

There can be a point of diminishing returns for the number of sensors located within a water system, although the budget is a limiting factor that may be exceeded first. In addition, Tryby et al. (2010) found that the number of monitoring stations in a network is more important than the method used to place them.

3.4.2. Integration of Human Health Factors

When applying sensor placement formulations, patterns regarding the use of drinking water need to be taken into account to more accurately represent health effects of a contamination event. Davis and Janke (2009) developed a probabilistic time-dependent model for the ingestion of tap water. Inclusion of consumption characteristics is important for understanding health effects of a contamination event, since it may be difficult to relate a public health emergency to the drinking water supply. For example, biological agents may be difficult to trace back to water contamination, because the agent may be present only briefly, incubation periods can vary, and individuals may become infected through contact with infected individuals (Murray et al., 2006).

Murray et al. (2006) described a dynamic disease transmission model that quantifies the spread of disease through population subgroups, including populations that are susceptible and have compromised immune systems. This model includes infection resulting from consumption of contaminated drinking water. The infectivity rate is a function of the amount of water consumed and contaminant concentration. In a simulated drinking water contamination event, more than 50% of exposures happen within the first 20 hours (Murray et al., 2006). It is typically not practical to respond quickly enough to a water threat to prevent exposure to a contaminant. This model can be used for determining the locations and number of sensors to use in detection, based on minimization of infection in the exposed population.

Design of a CWS based on estimates of health effects can be complicated, as a result of the various contaminants that could be used, nature of the injection, hydraulic characteristics in the water system, and varying population exposed to the contaminant. Different contaminants at varying concentrations have differing health effects and can affect a population differently. For example, Davis and Janke (2011) found that contaminants injected at high concentrations are more likely to have a concentrated effect near the source, while lower doses have a diffuse effect, affecting a larger geographical area.

3.4.3. Event Detection Systems

Once a CWS that includes online contaminant monitoring is in place, an event detection system (EDS) can be used to analyze the data sets collected using the sensors and determine if contamination is occurring. Using standard water quality and operating parameters as surrogates for contamination results in large data sets that require analysis to interpret the data and predict contamination. Event detection and identification is an inverse modeling problem because contamination injection characteristics are unknown (location, time, concentration), and typically more than one solution is possible. Solutions to potential contaminant locations are underdetermined problems with multiple possible solutions, making them difficult to solve (Tryby et al., 2010). Considerable computational power is needed, or the problem, as posed, requires simplification. In addition to knowing whether or not an event is taking place, it is useful to know the approximate location of contaminant injection as well as the time of injection, duration of the injection, contaminant concentration, total mass injected, and extent of distribution system affected by the contamination event.

Considerable effort has been expended to develop EDSs for online contaminant monitoring systems that function as sensor networks. Most of the approaches developed for EDSs rely on using water distribution models that simulate hydraulic and contaminant transport phenomena. The contaminant source is located using a variety of approaches. An optimization approach that involves determining all possible contaminant source locations is often used. Since the problem is underdefined due to the large number of system nodes and the small number of sensors deployed, sophisticated algorithms and considerable computing power are often needed to develop a set of reasonable source solutions. Various researchers have addressed the contaminant source problem using a variety of approaches.

CANARY Software

To assist with detecting contamination events, the U.S. EPA developed the event detection system software CANARY, which can be used to analyze data collected by a network of online contaminant sensors (U.S. EPA, 2009a, 2010b). This software was developed at Sandia National Laboratories in collaboration with U.S. EPA's National Homeland Security Research Center. It was written in the MATLAB® programming language, and it functions as a stand-alone software program, although it should be tied into the existing supervisory control and data acquisition system, so that the water utility operators can access the data and results. Third party software can be used to relay data between CANARY and SCADA systems (e.g., U.S. EPA's EDDIES software).

The basis of CANARY is that baseline water quality conditions are established, then the CANARY program is used to determine if the data collected by the monitors is within established tolerances or if an anomalous water quality event is occurring, suggesting that a threat may be present (Murray et al., 2010; U.S. EPA, 2010a). CANARY can be used in real time or it can be used to analyze previously collected data sets. Water quality changes within a distribution system can occur as a result of the following situations (U.S. EPA, 2010b):

- Baseline change: sudden and persistent change due to operational occurrence, such as turning on a pump station or closing a valve.
- Outlier: unexpected value that occurs for a very short period of time.
- Event: values are significantly different than expected for a minimum number of time steps (setting are adjustable by user).

Within CANARY, calculations are performed to determine the probability of an anomalous event based on the data collected as well as tolerances determined based on historical data sets. False alarms are a serious concern for water utilities setting up a CWS for better security. There are several ways to minimize false positives. Monitoring stations can be located in areas with stable water quality. For example, locating monitoring stations away from pumps, valves, and other features that cause unusual hydraulic characteristics may be helpful. In addition, algorithms can be incorporated into the software program to recognize normal and expected changes in water quality. Developers of the program state that, by properly setting up the CANARY software, false positives have been reduced by 65% for data sets that included simulated water quality events (U.S. EPA, 2010b).

Three algorithms within CANARY can be used to detect contamination events: the linear prediction-correction filter prediction algorithm, multivariate nearest neighbor prediction algorithm, and set-point proximity algorithm (U.S. EPA, 2009a, 2010b). A binomial event discriminator evaluates outliers within a specific time window to determine if an anomalous event has occurred. Selection of the time window will impact accuracy, but also cause false alarms. CANARY contains water quality pattern matching capability where historical data can be used to identify recurring patterns, although these patterns may not always occur at the same time of day. In using CANARY, different results can be obtained by adjusting the moving window that affects how data is pooled (Murray et al., 2012). A trajectory clustering approach can be used to identify patterns within the time series data sets and by identifying recurring patterns, the frequency of false alarms is reduced (U.S. EPA, 2009a).

Optimization Techniques

An optimization approach is often taken for solving the event detection problem (Murray et al., 2009). Murray et al. (2006) described a decision framework comprising a modeling process and a decision-making process that employs optimization. The modeling process included creating a network model for hydraulic and water quality analysis, describing sensor characteristics, defining the contamination threats, selecting performance measures, planning a utility response to sensor detection, and identifying potential sensor locations. The decision-making process involves applying an optimization method and evaluating sensor placements. The process is informed by analyzing trade-offs and comparing a series of designs to account for modeling and data uncertainties. This approach was applied to design a CWS for the first EPA Water Security Initiative (WSI) pilot city (U.S. EPA, 2005).

Various approaches have been used to solve optimization of event detection solutions (Ostfeld, 2006; Ostfeld and Salomons, 2005; Preis and Ostfeld, 2008b). Perelman et al. (2012) offered a comprehensive solution to event detection that relies on artificial neural networks with application of Bayes's rule and assessment of the model occurring using correlation coefficients, mean squared error, confusion matrices, receiver operating characteristic (ROC) curves, and true and false positive rates. Perelman et al. (2012) tested their approach on a real data set. Arad et al. (2013) followed up on the work of Perelman et al. (2012) by adding a dynamic threshold scheme. The decision variables were: positive and negative filters, positive and negative dynamic thresholds and data window size.

Previous approaches have focused on using EPANET with optimization formulation and genetic algorithms for solution of the optimization problem (Guidorzi et al., 2009). Slight modifications to this formulation are noted in many of the studies. In using a statistical model with heuristic search method, Liu et al. (2012) found that a local search is better for finding optimal solutions than a standard evolutionary algorithm. Di Cristo and Leopardi (2008) used selection of possible nodes, then the probability of each node being the contaminant source. False alarms are reduced using Kulldorff's scan test to determine if detections are significant (Koch and McKenna, 2011). Laird et al. (2006) used a two phase optimization formulation to first screen possible contamination sources. Preis and Ostfeld (2006) used a model tree linear programming approach to addressing EDS. Tryby et al. (2010) ran Monte Carlo simulations to determine the effectiveness of the sensor network to identify the location of contamination injection.

Researchers investigated the effect of imperfect sensors and low-resolution sensors on the ability of the sensor network to achieve event detection. Kumar et al. (2012) found that even low-resolution sensors with binary signals are useful. Although the low-resolution sensors result in more possible solutions to the contamination source problem, Kumar et al. (2012) found that the true source is contained within the set of possible solutions. Dawsey et al. (2006) used a Bayesian belief network (BBN) to develop an EDS method that is probabilistic and accounts for imperfect sensors. De Sanctis et al. (2010) used a particle back-tracking algorithm to identify possible contaminant sources, assuming that sensors provide only binary data. McKenna et al. (2006) investigated both the influence of the number of sensors and sensor resolution on the ability to detect contamination events.

Some of the event detection methodologies were tested using real water system data and simulated water threat events. Guan et al. (2006) tested an EDS formulation for Dover township, New Jersey, using a combination of historic hydraulic data and contamination events simulated in EPANET.

Receiver Operating Characteristics

Receiver operating characteristic (ROC) graphs are used to assist with data interpretation (U.S. EPA, 2007). ROC can be used for EDS to compare water quality changes with the risk of making the wrong decision (McKenna et al., 2008). A ROC curve is developed based on use of experimental data, where true contamination is compared with sensor data to define the true-positive rate as a function of the false-positive rate. Once established using real data sets, the ROC can be used for EDS (Kroll and King, 2010). McKenna et al. (2008) presented an analytical approach consisting of detection algorithms that allow analysis of water quality data to identify contamination events based on the development of ROC curves.

Data Mining

Data mining is an additional computational approach that can be used for EDS solution. Huang and McBean (2009) used a data mining approach with maximum likelihood for an EDS; a solution was determined quickly (within 3 minutes for a 285 node water system with five sensors) using this approach. Shen and McBean (2012) explored ways to reduce false-negatives and false-positives in event detection using a data mining approach with parallel computing to simulate scenarios.

Geographical Information Systems

GIS tools can be used to assist in managing data inputs from online sensors (Trepanier et al., 2006). An example is RiverSpill, a GIS-based contaminant transport software package that can be used to simulate contamination in source waters (Samuels et al., 2006).

3.4.4. Model Calibration and Validation

The water distribution system modeling software used as part of a CWS should be calibrated and validated using system data. Model calibration includes the following components: (1) definition of model variables, coefficients, and equations; (2) selection of the objective function; (3) selection of calibration data; and (4) selection of an optimization solution method (Ostfeld et al., 2012).

Distribution system modeling validation can be accomplished using desktop analysis to match hydraulic parameters, pressure studies, chlorine decay studies, and tracer studies (U.S. EPA, 2007). Savic et al. (2009) offered water distribution model calibration guidance. In 2010, the results of the Battle of the Water Calibration Networks were presented at a conference, an effort that gathered 14 teams to compare their results for a given problem (Ostfeld et al., 2012).

3.5. Studies Conducted in Distribution System Simulators

Several bench-scale and pilot-scale studies have been conducted to investigate use of online contaminant monitoring and the efficacy of sensors and software to detect contamination events. The pilot-scale systems consisted of recirculating pipeline networks that were referred to as distribution system simulators (DSSs). In most of these studies, a suite of water quality parameters was simultaneously monitored to indicate contamination.

Byer (2005) studied simulated contamination events by introducing the contaminants aldicarb, sodium arsenate, sodium cyanide, and sodium fluoroacetate into a bench-scale DSS. The water quality parameters measured were pH, chlorine residual, turbidity, and TOC (Byer, 2005). Using data from the sensors, Byer (2005) was able to detect three of the four chemicals at low concentrations, below that which would cause human health effects.

In the U.S. EPA study by Hall et al. (2007), the following standard water quality parameters were monitored: pH, free chlorine, ORP, dissolved oxygen, specific conductance (SC), turbidity, TOC, chloride, ammonia, and nitrate. The pilot-scale system was contaminated with the following: secondary effluent, potassium free cyanide, malathion insecticidal formulation, glyphosate herbicidal formulation, nicotine, arsenic trioxide, aldicarb, and *E. coli* with growth media. Hall et al. (2007) investigated the following sensor technologies:

- ATI: free chlorine.
- Hach Model A-15 Cl-17: free and total chlorine.
- Hach 1720 D: turbidity.
- GLI Model PHD: pH.
- GLI Model 3422: specific conductance.
- Hach Astro TOC UV process analyzer: TOC.
- Dascore Six-Sense Sonde: SC, DO, ORP, pH, temperature, free chlorine.
- YSI 6600 Sonde: SC, DO, ORP, pH, temperature, ammonia-nitrogen, chloride, nitrate-nitrogen, turbidity.
- Hydrolab Data Sonde 4a: SC, DO, ORP, pH, temperature, ammonia-nitrogen, chloride, nitrate-nitrogen, turbidity.

The results of Hall et al. (2007) indicated that no single sensor was capable of detecting all contamination events and the most effective sensors were those that measured free chlorine, TOC, ORP, specific conductance, and chloride. In an extension of the work conducted by the U.S. EPA DSS, Yang et al. (2009)

used the experimental data collected and tested EDS software using the following contaminants: pesticides, herbicides (aldicarb, glyphosate, and dicamba), nicotine, colchicine, *E. coli* in broth, biological growth media, mercuric chloride, and potassium ferricyanide. The water quality parameters tested were pH, ORP, free chlorine, total chlorine, and chloride (Yang et al., 2009).

In an investigation of detecting microbial contaminants as well as the particulates and organic compounds that might accompany microbial suspensions, the U.S. EPA (2010a) evaluated the following sensors:

- Fluid Imaging Technologies FlowCAM®, which uses a flow cytometer and microscope to count particles and collect information about the particle characteristics.
- Hach FilterTrak™ 660 sc Laser Nephelometer, which uses laser light for turbidity measurements.
- JMAR BioSentry®, which uses laser light with a multiple angle scattering device to classify particles as well as count them.
- Real UVT Online, which measures ultraviolet (UV) light transmission at a wave length of 254 nm.
- S::CAN Spectro::lyser™ and Carb::olyser™, which measure TOC, turbidity, and a host of other constituents, depending on how the unit is set up.

In their study, the U.S. EPA (2010a) found that online sensors were typically not able to detect microbial suspensions with concentrations lower than 10^5 cfu/ml. The JMAR BioSentry® was able to detect the lowest biological suspension concentration of 600 cfu/ml. In general, most of the sensors listed above performed as well as online sensors measuring free chlorine and TOC, which were able to detect microbial suspensions with concentrations as low as 2.5×10^4 cfu/ml. In addition to these sensors, the U.S. EPA tested the following sensors in DSSs in previous tests (U.S. EPA, 2009b):

- YSI 660
- YSI 6920DW
- Six-Cense™
- Hydrolab® DS5
- Troll® 9000
- Hach/GLI Model C53 Conductivity Analyzer
- ZAPS MP-1
- Hach CL17 Free Chlorine Analyzer
- Hach/GLI Model P53 pH/ORP Analyzer
- Hach 2200 PCX Particle Counter
- Hach AstroTOC™ UV Process Total Organic Carbon Analyzer
- Sievers® 900 On-Line Total Organic Carbon Analyzer
- Hach 1720D Turbidimeter

In the study by the U.S. EPA (2009b), many contaminants were evaluated, including the actual contaminants and surrogates of contamination (Table 5.3).

TABLE 5.3 Constituents Studied in a Pilot Study by the U.S. EPA (U.S. EPA, 2009b)

Contaminant Categories	Contaminants Used
Biological	*Bacillus globigii*, Bacteriophage MS2, *Escherichia coli*, surrogate beads
Insecticides	Aldicarb, nicotine, Real Kill®/malathion, dichlorvos, phorate
Herbicides	Roundup®/glyphosate, dicamba
Culture broths	Nutrient broth, sporulation media, Terrific broth, tryptic soy broth
Inorganics	Arsenic trioxide, cesium chloride, cobalt chloride, lead nitrate, mercuric chloride, potassium cyanide, potassium ferricyanide, sodium arsenite, sodium thiosulfate, sodium fluoride
Warfare agents	Ricin, G-type nerve agent, V-series nerve agent, potassium cyanide
Others	Blank (GAC water), secondary effluent, colchicine, dimethyl sulfoxide, dye Sucrose, sodium fluoroacetate, methanol

The choice of contaminants included in the study is indicative of the contaminants of concern. Similar to other study results, it was determined that free chlorine and TOC were the best indicators of contamination events.

Microbial suspensions are likely to exert a chlorine demand, making chlorine residual an effective water quality parameter for detecting microbial suspensions (Helbling and VanBriesen, 2008). Helbling and VanBriesen (2008) found that chlorine residual is a suitable indicator for bacterial intrusions in a DSS, although they found that fairly high bacterial concentrations of 10^5 CFU/ml, 10^4 CFU/ml, and 10^3 CFU/ml are required for *Escherichia coli*, *Staphylococcus epidermidis*, and *Mycobacterium aurum*, respectively, to incur a chlorine demand in the suspensions of these organisms. Chlorine residual in distribution systems following a contamination event was modeled by Helbling and VanBriesen (2009).

Fluorescence-based monitoring devices, which are more frequently used to detect hydrocarbons and algae, have been tested in DSSs to determine if these sensors outperform traditional water quality parameter sensors (U.S. EPA, 2012a). Advances in light emitting diode (LED) technology have made these devices more attainable. The two units tested by the U.S. EPA (2012a) were the Turner Designs (Turner Designs Hydrocarbon Instruments Inc.) Model

TD1000C, which uses a single excitation and emission wavelength, and the ZAPS LiquID™ unit (ZAPS Technologies Inc.), which uses multiple wavelengths. Although the fluorescence-based devices performed reasonably well in detecting contaminants in simulated events, the fluorometers were not as effective as multiparameter water quality monitors. The U.S. EPA (2012a) found that free chlorine and TOC were the most responsive parameters for detecting a wide range of contamination events. One of the main concerns about the fluorescence-based devices was the cost; in addition to having higher capital costs, the units require more oversight and maintenance because of the optical sensors (U.S. EPA, 2012a).

The U.S. EPA study in a DSS was repeated to study differences between chlorinated and chloraminated drinking water. Experimental work indicates that it may be more difficult to conduct contaminant monitoring in chloraminated systems than in systems that use free chlorine as a disinfectant residual, because chlorine residual monitors do not work as well in chloraminated waters (Szabo et al., 2008). In a similar study by Hall et al. (2007), where chlorinated water was used, chlorine residual was one of the more effective water quality parameters for indicating contamination.

Installation of online contaminant monitoring requires a capital investment as well as continual costs to keep these systems maintained and functional. In the U.S. EPA (2010a) study of online sensors used to detect biological contaminants, the costs of the sensors ranged from $5,000 to $50,000 for the sensors alone, which did not include installation costs or costs for ancillary systems, such as electrical, plumbing, and communication systems. Costs may also be incurred to physically house the sensor and related infrastructure. The costs of installing and maintaining sensors may be especially high if the sensors are located in remote areas and in areas distant from other infrastructure. Maintenance of online monitoring stations includes replacing reagents, calibration, cleaning tubing, and cleaning electrodes. In the study by Hall et al. (2007), most sensors had weekly or monthly calibration schedules.

3.6. Case Study of a Contaminant Warning System

As part of the Water Security Initiative, the U.S. EPA sponsored a pilot demonstration study for a multifaceted contaminant warning system (CWS) in a full-scale water system (U.S. EPA, 2008a). The project took place at the Greater Cincinnati Water Works system in Cincinnati, Ohio, and occurred over two years, from December 2005 to December 2007 (Fencil and Hartman, 2009). During the pilot study multiple components of a CWS were developed for the city, including contaminant monitoring and sensor network systems. The CWS included 17 multiprobe online monitoring stations, located throughout the water distribution system. The sensors were used to measure free chlorine, TOC, ORP, SC, pH, and turbidity. Two EDS software programs were installed and tested. A dedicated communications network was used for the project, with online water quality data being transmitted to the SCADA system and the EDS programs. The U.S. EPA EDDIES program was used to interface with the SCADA system and relay data to CANARY (Murray et al., 2010). The other EDS software used were Event Monitor by Hach, con::stat by S::can, and BlueBox by Whitewater Security. Data transmission occurred by a digital cellular network.

In addition to installing and testing an online contaminant monitoring system, the pilot project resulted in the water utility establishing other aspects of a complete CWS. As a result of the project, a laboratory network capable of detecting 10 out of 12 high-priority contaminant classes was established (U.S. EPA, 2008a). Field screening and site characterization procedures were developed, involving local hazmat teams and resulting in field sampling kits being assembled. The physical security equipment installed consisted of video cameras, motion sensors, door contact switches at three large pump stations, ladder motion sensors at seven elevated storage tanks, and vent housings installed at reservoirs and tanks. In addition, a dedicated SCADA system was installed for the physical security of system components. The water utility implemented a centralized consumer complaint system with electronic data management and automated data analysis of consumer complaint data. Systems were set up to record public health data and communications were improved with local public health agencies. The water utility developed contaminant incident response plans that resulted in communication with their response partners. Training and exercises were implemented to practice the emergency response procedures. The total cost of the pilot CWS was $12.3 million (U.S. EPA, 2008a).

During the initial implementation of the CWS, there were a lot of false alarms, up to 10 per day (Murray et al., 2010). Most of the alarms were due to sensor problems (approximately two thirds of all alarms); bad data caused by interruptions in communications, equipment calibration events, operational changes, and system water quality events caused the remainder of the alarms (Murray et al., 2010). In some cases, the water quality events occurred for unknown reasons. Some of the problems experienced with the sensors included depletion of reagents, clogged tubing, syringe pump failure, and electrodes requiring cleaning or wearing out (Murray et al., 2010). Operational events that caused significant changes in water quality measured by the online monitors included switching of tanks and reservoirs, valves opening and closing, carbon recharge, pH control issues, and changes in chlorine dosage at the treatment plant (Murray et al., 2010).

The incidences of false alarms were reduced over time by optimizing the system settings (Allgeier et al., 2011; Pickard et al., 2011). One of the elements that affected whether or not an alarm was triggered was selection of the history window in CANARY, which basically directs the program regarding how the data are pooled. To reduce the occurrence of false alarms, the CANARY window was adjusted. Other examples of changes made to the CWS to reduce alarms included replacing motion sensors with hatch switches, programming delays into ladder alerts, and changing the public health alert criteria. As a result of the optimization, total system false alarms were reduced by 87% (Allgeier et al., 2011). Following implementation of the Cincinnati CWS, further analysis was completed to achieve additional optimization (Pickard et al., 2011). Reviewers of the system focused on system availability and data completeness. The system periodically experienced downtime for a variety of reasons. By incorporating redundancy into the CWS, the effects of downtime became negligible (Pickard et al., 2011).

Once the Cincinnati CWS was online, a full-scale exercise was developed to simulate a water threat and allow the water utility personnel to practice the emergency response protocol and consequence management training (Fencil and Hartman, 2009). The practice exercise included interacting with other agencies and allowed water system personnel to practice the emergency response protocols. During the exercise, a portion of the distribution system was effectively isolated.

CONCLUSIONS

Detection of contaminants in drinking water systems is a critical component of a defense-in-depth approach to water security. Detection necessitates routine monitoring of water quality constituents and contaminants as well as a systematic process for identifying anomalies. Existing water quality monitoring strategies were developed to meet regulatory drinking water standards rather than protect against water contamination events. Currently, the detection of specific chemicals and biological contaminants that could be used in an intentional contamination event generally requires use of advanced analytical methods. In most cases, these analyses may require sending samples to outside laboratories, further increasing detection time and delaying a response. Advances in real-time detection technology, including on-site test kits, biosensors, and online sensors, have facilitated contaminant detection but are still limited by their range of detectable contaminants. Despite the limitations, real-time detection is vital, as it provides the basis for a quick response and rehabilitation plan, which limits the overall impact of the contamination event.

Despite ongoing research, the detection of contaminants in drinking water systems is still limited by capability of current technology and the economic burden placed on the water utility to deploy such systems. The economic burden can be mitigated by the implementation of dual-use detection technologies that provide additional benefits. However, further research is needed to expand the detection capabilities, reduce the cost, and optimize the placement of sensors to make real-time detection sensors more accessible, comprehensive, and widespread in the nation's water systems. Also, additional field testing of detection sensors and software is needed to validate these technology solutions.

REFERENCES

Afshar, A., Marino, M.A., 2012. Multi-objective coverage-based ACO model for quality monitoring in large water networks. Water Resour. Manag. 26 (8), 2159–2176.

Ailamaki, A., Faloutsos, C., Fischbeck, P.S., Small, M.J., VanBriesen, J., 2003. An environmental sensor network to determine drinking water quality and security. Sigmod Record 32 (4), 47–52.

Aisopou, A., Stoianov, I., Graham, N.J.D., 2012. In-pipe water quality monitoring in water supply systems under steady and unsteady state flow conditions: A quantitative assessment. Water Res. 46 (1), 235–246.

Alfonso, L., Jonoski, A., Solomatine, D., 2010. Multiobjective optimization of operational responses for contaminant flushing in water distribution networks. J. Water Resour. Plann. Manage. ASCE 136 (1), 48–58.

Allgeier, S.C., Haas, A.J., Pickard, B.C., 2011. Optimizing alert occurrence in the Cincinnati contamination warning system. J. Am. Water Works Assoc. 103 (10), 55–66.

Arad, J., Housh, M., Perelman, L., Ostfeld, A., 2013. A dynamic thresholds scheme for contaminant event detection in water distribution systems. Water Res. 47 (5), 1899–1908.

Berry, J., Carr, R.D., Hart, W.E., Leung, V.J., Phillips, C.A., Watson, J.P., 2009. Designing contamination warning systems for municipal water networks using imperfect sensors. J. Water Resour. Plann. and Manag. ASCE 135 (4), 253–263.

Berry, J., Hart, W.E., Phillips, C.A., Uber, J.G., Watson, J.P., 2006. Sensor placement in municipal water networks with temporal integer programming models. J. Water Resour. Plann. Manag. ASCE 132 (4), 218–224.

Berry, J.W., Fleischer, L., Hart, W.E., Phillips, C.A., Watson, J.P., 2005. Sensor placement in municipal water networks. J. Water Resour. Plann. and Manag. ASCE 131 (3), 237–243.

Botsford, J.L., 2002. A comparison of ecotoxicological tests. Atla-Alternatives to Laboratory Animals 30 (5), 539–550.

Bukhari, Z., Weihe, J.R., Lechevallier, M.W., 2007. Rapid detection of *Escherichia coli* O157:H7 in water. J. Am. Water Works Assoc. 99 (9), 157–167.

Byer, D., 2005. Real-time detection of intentional chemical contamination–In the distribution system. J. Am. Water Works Assoc. 97 (7), 130–141.

Campbell, C.G., Mascetti, M.M., Hoppes, W., Stringfellow, W.T., 2007. Measurement reproducibility of the Bioscan™ flow-through respirometer applied as a toxicity-based early warning system for water contamination. Environ. Pract. 9 (1), 42–53.

Chang, I.H., Tulock, J.J., Liu, J.W., Kim, W.S., Cannon, D.M., Lu, Y., Bohn, P.W., Sweedler, J.V., Cropek, D.M., 2005. Miniaturized lead sensor based on lead-specific DNAzyme in a nanocapillary interconnected microfluidic device. Environ. Sci. Technol. 39 (10), 3756–3761.

Chang, N.B., Pongsanone, N.P., Ernest, A., 2011. Comparisons between a rule-based expert system and optimization models for sensor deployment in a small drinking water network. Expert Systems with Applications 38 (8), 10685–10695.

Cozzolino, L., Della Morte, R., Palumbo, A., Pianese, D., 2011. Stochastic approaches for sensors placement against intentional contaminations in water distribution systems. Civil Eng. Environ. Sys. 28 (1), 75–98.

Crisologo, J., 2008. California implements water security and emergency preparedness, response, and recovery initiatives. J. Am. Water Works Assoc. 100 (7), 30–34.

Curtis, T.M., Tabb, J., Romeo, L., Schwager, S.J., Widder, M.W., van der Schalie, W.H., 2009a. Improved cell sensitivity and longevity in a rapid impedance-based toxicity sensor. J. Appl. Toxicol. 29 (5), 374–380.

Curtis, T.M., Widder, M.W., Brennan, L.M., Schwager, S.J., van der Schalie, W.H., Fey, J., Salazar, N., 2009b. A portable cell-based impedance sensor for toxicity testing of drinking water. Lab on a Chip 9 (15), 2176–2183.

Davila, D., Esquivel, J.P., Sabate, N., Mas, J., 2011. Silicon-based microfabricated microbial fuel cell toxicity sensor. Biosensors Bioelectronics 26 (5), 2426–2430.

Davis, M.J., Janke, R., 2009. Development of a probabilistic timing model for the ingestion of tap water. J. Water Resour. Plann. and Manag. ASCE 135 (5), 397–405.

Davis, M.J., Janke, R., 2011. Patterns in potential impacts associated with contamination events in water distribution systems. J. Water Resour. Plann. Manag. ASCE 137 (1), 1–9.

Dawsey, W.J., Minsker, B.S., VanBlaricum, V.L., 2006. Bayesian belief networks to integrate monitoring evidence of water distribution system contamination. J. Water Resour. Plann. Manag. ASCE 132 (4), 234–241.

de Hoogh, C.J., Wagenvoort, A.J., Jonker, F., van Leerdam, J.A., Hogenboom, A.C., 2006. HPLC-DAD and Q-TOF MS techniques identify cause of *Daphnia* biomonitor alarms in the River Meuse. Environ. Sci. Technol. 40 (8), 2678–2685.

de Mora, K., Joshi, N., Balint, B.L., Ward, F.B., Elfick, A., French, C.E., 2011. A pH-based biosensor for detection of arsenic in drinking water. Anal. Bioanal. Chem. 400 (4), 1031–1039.

De Sanctis, A.E., Shang, F., Uber, J.G., 2010. Real-time identification of possible contamination sources using network backtracking methods. J. Water Resour. Plann. Manag. ASCE 136 (4), 444–453.

Deng, L., Chen, C.G., Zhou, M., Guo, S.J., Wang, E.K., Dong, S.J., 2010. Integrated self-powered microchip biosensor for endogenous biological cyanide. Anal. Chem. 82 (10), 4283–4287.

Deshpande, K., Mishra, R.K., Bhand, S., 2010. A high sensitivity micro format chemiluminescence enzyme inhibition assay for determination of Hg(II). Sensors 10 (7), 6377–6394.

Dewhurst, R.E., Wheeler, J.R., Chummun, K.S., Mather, J.D., Callaghan, A., Crane, M., 2002. The comparison of rapid bioassays for the assessment of urban groundwater quality. Chemosphere 47 (5), 547–554.

Di Cristo, C., Leopardi, A., 2008. Pollution source identification of accidental contamination in water distribution networks. J. Water Resour. Plann. Manag. ASCE 134 (2), 197–202.

Dierksen, K.P., Mojovic, L., Caldwell, B.A., Preston, R.R., Upson, R., Lawrence, J., McFadden, P.N., Trempy, J.E., 2004. Responses of fish chromatophore-based cytosensor to a broad range of biological agents. J. Appl. Toxicol. 24 (5), 363–369.

Dorini, G., Jonkergouw, P., Kapelan, Z., Savic, D., 2010. SLOTS: Effective algorithm for sensor placement in water distribution systems. J. Water Resour. Plann. Manag. ASCE 136 (6), 620–628.

Eliades, D.G., Polycarpou, M.M., 2010. A fault diagnosis and security framework for water systems. IEEE Trans. Control Sys. Techn. TCST-18 (6), 1254–1265.

Eltzov, E., Marks, R.S., 2010. Fiber-optic based cell sensors. Adv. Biochem. Eng. Biotechnol. 117, 131–154.

Eltzov, E., Marks, R.S., 2011. Whole-cell aquatic biosensors. Anal. Bioanal. Chem. 400 (4), 895–913.

Ercole, C., Del Gallo, M., Pantalone, M., Santucci, S., Mosiello, L., Laconi, C., Lepidi, A., 2002. A biosensor for *Escherichia coli* based on a potentiometric alternating biosensing (PAB) transducer. Sensors and Actuators B—Chemical 83 (1–3), 48–52.

Fencil, J., Hartman, D., 2009. Cincinnati's drinking water Contamination Warning System goes through full-scale exercise. J. Am. Water Works Assoc. 101 (2), 52–55.

Foran, J.A., Brosnan, T.M., 2000. Early warning systems for hazardous biological agents in potable water. Environ. Health Perspect. 108 (10), 993–996.

Gardeniers, H., Van den Berg, A., 2004. Micro- and nanofluidic devices for environmental and biomedical applications. Int. J. Environ. Anal. Chem. 84 (11), 809–819.

Geng, P., Zhang, X.A., Teng, Y.Q., Fu, Y., Xu, L.L., Xu, M., Jin, L.T., Zhang, W., 2011. A DNA sequence-specific electrochemical biosensor based on alginic acid-coated cobalt magnetic beads for the detection of E. coli. Biosensors Bioelectronics 26 (7), 3325–3330.

Giaever, I., Keese, C.R., 1993. A morphological biosensor for mammalian cells. Nature 366 (6455), 591–592.

Grayman, W.M., Ostfeld, A., Salomons, E., 2006. Locating monitors in water distribution systems: Red team–blue team exercise. J. Water Resour. Plann. Manag. ASCE 132 (4), 300–304.

Grimmett, P.E., Munch, J.W., 2013. Development of EPA Method 525.3 for the analysis of semi-volatiles in drinking water. Anal. Methods 5 (1), 151–163.

Groves, W.A., Grey, A.B., O'Shaughnessy, P.T., 2006. Surface acoustic wave (SAW) microsensor array for measuring VOCs in drinking water. J. Environ. Monit. 8 (9), 932–941.

Guan, J.B., Aral, M.M., Maslia, M.L., Grayman, W.M., 2006. Identification of contaminant sources in water distribution systems using simulation-optimization method: Case study. J. Water Resour. Plann. Manag. ASCE 132 (4), 252–262.

Guidorzi, M., Franchini, M., Alvisi, S., 2009. A multi-objective approach for detecting and responding to accidental and intentional contamination events in water distribution systems. Urban Water J. 6 (2), 115–135.

Hall, J., Zaffiro, A.D., Marx, R.B., Kefauver, P.C., Krishnan, E.R., Herrmann, J.G., 2007. On-line water quality parameters as indicators of distribution system contamination. J. Am. Water Works Assoc. 99 (1), 66–77.

Hart, W.E., Murray, R., 2010. Review of sensor placement strategies for contamination warning systems in drinking water distribution systems. J. Water Resour. Plann. Manag. ASCE 136 (6), 611–619.

Helbling, D.E., VanBriesen, J.M., 2007. Free chlorine demand and cell survival of microbial suspensions. Water Res. 41 (19), 4424–4434.

Helbling, D.E., VanBriesen, J.M., 2008. Continuous monitoring of residual chlorine concentrations in response to controlled microbial intrusions in a laboratory-scale distribution system. Water Res. 42 (12), 3162–3172.

Helbling, D.E., VanBriesen, J.M., 2009. Modeling residual chlorine response to a microbial contamination event in drinking water distribution systems. J. Environ. Eng. ASCE 135 (10), 918–927.

Hillaker, T.L., Botsford, J.L., 2004. Toxicity of herbicides determined with a microbial test. Bull. Environ. Contam. Toxicol. 73 (3), 599–606.

Ho, C.K., Robinson, A., Miller, D.R., Davis, M.J., 2005. Overview of sensors and needs for environmental monitoring. Sensors 5 (1–2), 4–37.

Hrudey, S.E., Rizak, S., 2004. Discussion of "Rapid analytical techniques for drinking water security investigations". J. Am. Water Works Assoc. 96 (9), 110–113.

Huang, J.J., McBean, E.A., 2009. Data mining to identify contaminant event locations in water distribution systems. J. Water Resour. Plann. Manag. ASCE 135 (6), 466–474.

Isovitsch, S.L., VanBriesen, J.M., 2008. Sensor placement and optimization criteria dependencies in a water distribution system. J. Water Resour. Plann. Manag. ASCE 134 (12), 186–196.

Iuga, A., Lerner, E., Shedd, T., Van der Schalie, W.H., 2009. Rapid responses of a melanphore cell line to chemical contaminants in water. J. Appl. Toxicol. 29, 346–349.

Jang, A., Zou, Z.W., Lee, K.K., Ahn, C.H., Bishop, P.L., 2011. State-of-the-art lab chip sensors for environmental water monitoring. Measurement Sci. Technol. 22 (3).

Janke, R., Murray, R., Uber, J., Taxon, T., 2006. Comparison of physical sampling and real-time monitoring strategies for designing a contamination warning system in a drinking water distribution system. J. Water Resour. Plann. Manag. ASCE 132 (4), 310–313.

Janke, R.J., Davis, M., Taxon, T., 2011. Simulating intentional contamination events in water distribution systems: A report on the sensitivity of estimated impacts to major simulation parameters. In: Proceedings of the 12th Annual Conference on Water Distribution Systems Analysis (WDSA). September 12–15, 2010. Tucson, AZ, pp. 568–584.

Jeon, J., Kim, J.H., Lee, B.C., Kim, S.D., 2008. Development of a new biomonitoring method to detect the abnormal activity of *Daphnia magna* using automated Grid Counter device. Sci. Total Environ. 389 (2–3), 545–556.

Joshi, N., Wang, X., Montgomery, L., Elfick, A., French, C.E., 2009. Novel approaches to biosensors for detection of arsenic in drinking water. Desalination 248 (1–3), 517–523.

Kessler, A., Ostfeld, A., Sinai, G., 1998. Detecting accidental contaminations in municipal water networks. J. Water Resour. Plann. Manag. ASCE 124 (4), 192–198.

Khanal, N., Buchberger, S.G., McKenna, S.A., 2006. Distribution system contamination events: Exposure, influence, and sensitivity. J. Water Resour. Plann. Manag. ASCE 132 (4), 283–292.

Koch, M.W., McKenna, S.A., 2011. Distributed sensor fusion in water quality event detection. J. Water Resour. Plann. Manag. ASCE 137 (1), 10–19.

Krause, A., Leskovec, J., Guestrin, C., VanBriesen, J., Faloutsos, C., 2008. Efficient sensor placement optimization for securing large water distribution networks. J. Water Resour. Plann. Manag. ASCE 134 (6), 516–526.

Kroll, D., King, K., 2010. Methods for evaluating water distribution network early warning systems. J. Am. Water Works Assoc. 102 (1), 79–89.

Kumar, J., Brill, E.D., Mahinthakumar, G., Ranjithan, S.R., 2012. Contaminant source characterization in water distribution systems using binary signals. J. Hydroinformatics 14 (3), 585–602.

Laird, C.D., Biegler, L.T., Waanders, B., 2006. Mixed-integer approach for obtaining unique solutions in source inversion of water networks. J. Water Resour. Plann. Manag. ASCE 132 (4), 242–251.

Lewis, P.R., Manginell, R.P., Adkins, D.R., Kottenstette, R.J., Wheeler, D.R., Sokolowski, S.S., Trudell, D.E., Byrnes, J.E., Okandan, M., Bauer, J.M., Manley, R.G., Frye-Mason, G.C., 2006. Recent advancements in the gas-phase MicroChemLab. IEEE Sensors J 6 (3), 784–795.

Lindquist, H.D.A., Harris, S., Lucas, S., Hartzel, M., Riner, D., Rochele, P., DeLeon, R., 2007. Using ultrafiltration to concentrate and detect *Bacillus anthracis, Bacillus atrophaeus* subspecies globigii, and *Cryptosporidium parvum* in 100-liter water samples. J. Microbiol. Methods 70 (3), 484–492.

Liu, L., Zechman, E.M., Mahinthakumar, G., Ranjithan, S.R., 2012. Coupling of logistic regression analysis and local search methods for characterization of water distribution system contaminant source. Eng. Appl. Artif. Intell. 25 (2), 309–316.

Magnuson, M.L., Allgeier, S.C., Koch, B., De Leon, R., Hunsinger, R., 2005. Responding to water contamination threats. Environ. Sci. Technol. 39 (7), 153A.

McKenna, S.A., Hart, D.B., Yarrington, L., 2006. Impact of sensor detection limits on protecting water distribution systems from contamination events. J. Water Resour. Plann. Manag. ASCE 132 (4), 305–309.

McKenna, S.A., Wilson, M., Klise, K.A., 2008. Detecting changes in water quality data. J. Am. Water Works Assoc. 100 (1), 74–85.

Murray, R., Hart, W.E., Phillips, C.A., Berry, J., Boman, E.G., Carr, R.D., Riesen, L.A., Watson, J.P., Haxton, T., Herrmann, J.G., Janke, R., Gray, G., Taxon, T., Uber, J.G., Morley, K.M., 2009. US Environmental Protection Agency uses operations research to reduce contamination risks in drinking water. Interfaces 39 (1), 57–68.

Murray, R., Haxton, T.M., McKenna, S., Hart, D., Umberg, K., Hall, J., Lee, Y., Tyree, M., Hartman, D., 2010. Case study application of the Canary event detection software. American Water Works Association (AWWWA) Water Quality Technology Conference and Exposition (WQTC). AWWA, Savannah, GA.

Murray, R., Janke, R., Hart, W.E., Berry, J.W., Taxon, T., Uber, J., 2008. Sensor network design of contamination warning systems: A decision framework. J. Am. Water Works Assoc. 100 (1), 97–109.

Murray, R., Uber, J., Janke, R., 2006. Model for estimating acute health impacts from consumption of contaminated drinking water. J. Water Resour. Plann. Manage. ASCE 132 (4), 293–299.

Murray, S., Ghazali, M., McBean, E.A., 2012. Real-time water quality monitoring: assessment of multisensor data using bayesian belief networks. J. Water Resour. Plann. Manag. ASCE 138 (1), 63–70.

Ostfeld, A., 2006. Enhancing water-distribution system security through modeling. J. Water Resour. Plann. Manag. ASCE 132 (4), 209–210.

Ostfeld, A., Salomons, E., 2004. Optimal layout of early warning detection stations for water distribution systems security. J. Water Resour. Plann. Manag. ASCE 130 (5), 377–385.

Ostfeld, A., Salomons, E., 2005. Securing water distribution systems using online contamination monitoring. J. Water Resour. Plann. Manag. ASCE 131 (5), 402–405.

Ostfeld, A., Salomons, E., Ormsbee, L., Uber, J.G., Bros, C.M., Kalungi, P., Burd, R., Zazula-Coetzee, B., Belrain, T., Kang, D., Lansey, K., Shen, H.L., McBean, E., Wu, Z.Y., Walski, T., Alvisi, S., Franchini, M., Johnson, J.P., Ghimire, S.R., Barkdoll, B.D., Koppel, T., Vassiljev, A., Kim, J.H., Chung, G.H., Yoo, D.G., Diao, K.G., Zhou, Y.W., Li, J., Liu, Z.L., Chang, K., Gao, J.L., Qu, S.J., Yuan, Y.X., Prasad, T.D., Laucelli, D., Lyroudia, L.S.V., Kapelan, Z., Savic, D., Berardi, L., Barbaro, G., Giustolisi, O., Asadzadeh, M., Tolson, B.A., McKillop, R., 2012. Battle of the water calibration networks. J. Water Resour. Plann. Manag. ASCE 138 (5), 523–532.

Ostfeld, A., Uber, J.G., Salomons, E., Berry, J.W., Hart, W.E., Phillips, C.A., Watson, J.P., Dorini, G., Jonkergouw, P., Kapelan, Z., di Pierro, F., Khu, S.T., Savic, D., Eliades, D., Polycarpou, M., Ghimire, S.R., Barkdoll, B.D., Gueli, R., Huang, J.J., McBean, E.A., James, W., Krause, A., Leskovec, J., Isovitsch, S., Xu, J.H., Guestrin, C., VanBriesen, J., Small, M., Fischbeck, P., Preis, A., Propato, M., Piller, O., Trachtman, G.B., Wu, Z.Y., Walski, T., 2008. The battle of the water sensor networks (BWSN): A design challenge for engineers and algorithms. J. Water Resour. Plann. Manag. ASCE 134 (6), 556–568.

Perelman, L., Arad, J., Housh, M., Ostfeld, A., 2012. Event detection in water distribution systems from multivariate water quality time series. Environ. Sci. Technol. 46 (15), 8212–8219.

Perelman, L., Ostfeld, A., 2010. Extreme impact contamination events sampling for water distribution systems security. J. Water Resour. Plann. Manage. ASCE 136 (1), 80–87.

Pickard, B.C., Haas, A.J., Allgeier, S.C., 2011. Optimizing operational reliability of the Cincinnati contamination warning system. J. Am. Water Works Assoc. 103 (1), 60–68.

Porco, J.W., 2010. Municipal water distribution system security study: Recommendations for science and technology investments. J. Am. Water Works Assoc. 102 (4), 30–32.

Preis, A., Ostfeld, A., 2006. Contamination source identification in water systems: A hybrid model trees-linear programming scheme. J. Water Resour. Plann. Manag. ASCE 132 (4), 263–273.

Preis, A., Ostfeld, A., 2008a. Genetic algorithm for contaminant source characterization using imperfect sensors. Civil Eng. Environ. Sys. 25 (1), 29–39.

Preis, A., Ostfeld, A., 2008b. Multiobjective contaminant response modeling for water distribution systems security. J. Hydroinformatics 10 (4), 267–274.

Preis, A., Ostfeld, A., 2008c. Multiobjective contaminant sensor network design for water distribution systems. J. Water Resour. Plann. Manage. ASCE 134 (4), 366–377.

Propato, M., 2006. Contamination warning in water networks: General mixed-integer linear models for sensor location design. J. Water Resour. Plann. Manage. ASCE 132 (4), 225–233.

Radix, P., Leonard, M., Papantoniou, C., Roman, G., Saouter, E., Gallotti-Schmitt, S., Thiebaud, H., Vasseur, P., 2000. Comparison of four chronic toxicity tests using algae, bacteria, and invertebrates assessed with sixteen chemicals. Ecotoxicol. Environ. Saf. 47 (2), 186–194.

Rico-Ramirez, V., Frausto-Hernandez, S., Diwekar, U.M., Hernandez-Castro, S., 2007. Water networks security: A two-stage mixed-integer stochastic program for sensor placement under uncertainty. Comp. Chem. Eng. 31 (5–6), 565–573.

Roberson, J.A., Morley, K.M., 2005. Contamination warning systems for water: An approach for providing actionable information to decision-makers. AWWA, Denver, CO.

Samuels, W.B., Amstutz, D.E., Bahadur, R., Pickus, J.M., 2006. RiverSpill: A national application for drinking water protection. J. Hydraulic Eng. ASCE 132 (4), 393–403.

Samuels, W.B., Bahadur, R., 2006. An integrated water quality security system for emergency response. Security of water supply systems: From source to tap. NATO Security through Science Series 8, 99–112.

Savic, D.A., Kapelan, Z.S., Jonkergouw, P.M.R., 2009. Quo vadis water distribution model calibration? Urban Water J 6 (1), 3–22.

Sengupta, A., Mujacic, M., Davis, E.J., 2006. Detection of bacteria by surface-enhanced Raman spectroscopy. Anal. Bioanal. Chem. 386 (5), 1379–1386.

Serjeantson, B., McKenny, S., van Buskirk, R., 2011. Leverage operations data and improve utility performance. AWWA Opflow 37 (2), 10–15.

Serra, B., Morales, M.D., Zhang, J.B., Reviejo, A.J., Hall, E.H., Pingarron, J.M., 2005. In-a-day electrochemical detection of coliforms in drinking water using a tyrosinase composite biosensor. Anal. Chem. 77 (24), 8115–8121.

Shen, H., McBean, E., 2012. False negative/positive issues in contaminant source identification for water-distribution systems. J. Water Resour. Plann. Manag. ASCE 138 (3), 230–236.

Shen, H.L., McBean, E., 2011. Pareto optimality for sensor placements in a water distribution system. J. Water Resour. Plann. Manag. ASCE 137 (3), 243–248.

Shoji, R., Sakai, Y., Sakoda, A., Suzuki, M., 2000. Development of a rapid and sensitive bioassay device using human cells immobilized in macroporous microcarriers for the on-site evaluation of environmental waters. Appl. Microbiol. Biotechnol. 54 (3), 432–438.

Skadsen, J., Janke, R., Grayman, W., Samuels, W., TenBroek, M., Steglitz, B., Bahl, S., 2008. Distribution system on-line monitoring for detecting contamination and water quality changes. J. Am. Water Works Assoc. 100 (7), 81–94.

Skolicki, Z., Wadda, M.M., Houck, M.H., Arciszewski, T., 2006. Reduction of physical threats to water distribution systems. J. Water Resour. Plann. Manag. ASCE 132 (4), 211–217.

Smeti, E.M., Thanasoulias, N.C., Lytras, E.S., Tzoumerkas, P.C., Golfinopoulos, S.K., 2009. Treated water quality assurance and description of distribution networks by multivariate chemometrics. Water Res. 43 (18), 4676–4684.

States, S., Newberry, J., Wichterman, J., Kuchta, J., Scheuring, M., Casson, L., 2004. Rapid analytical techniques for drinking water security investigations. J. Am. Water Works Assoc. 96 (1), 52–64.

States, S., Scheuring, M., Kuchta, J., Newberry, J., Casson, L., 2003. Utility-based analytical methods to ensure public water supply security. J. Am. Water Works Assoc. 95 (4), 103–115.

States, S., Wichterman, J., Cyprych, G., Kuchta, J., Casson, L., 2006. A field sample concentration method for rapid response to security incidents. J. Am. Water Works Assoc. 98 (4), 115–121.

Storey, M.V., van der Gaag, B., Burns, B.P., 2011. Advances in on-line drinking water quality monitoring and early warning systems. Water Res. 45 (2), 741–747.

Szabo, J.G., Hall, J.S., Meiners, G., 2008. Sensor response to contamination in chloraminated drinking water. J. Am. Water Works Assoc. 100 (4), 33–40.

Trepanier, M., Gauthier, V., Besner, M.C., Prevost, M., 2006. A GIS-based tool for distribution system data integration and analysis. J. Hydroinformatics 8 (1), 13–24.

Tryby, M.E., Propato, M., Ranjithan, S.R., 2010. Monitoring design for source identification in water distribution systems. J Water Resources Planning and Management ASCE. 136 (6), 637–646.

U.S. Environmental Protection Agency (U.S. EPA), 2003a. Module 3: Site characterization and sampling guide, EPA-817-D-03–003. Response Protocol Toolbox (RPTB) interim final: Planning for and responding to contamination threats to drinking water systems. EPA, Washington, DC.

U.S. Environmental Protection Agency (U.S. EPA), 2003b. Module 4: Analytical guide, EPA-817-D-03–004. Response Protocol Toolbox (RPTB) interim final: Planning for and responding to contamination threats to drinking water systems. EPA, Washington, DC.

U.S. Environmental Protection Agency (U.S. EPA), 2005. Technologies and techniques for early warning systems to monitor and evaluate drinking water quality: A state-of-the-art review, EPA/600/R-05/156. EPA, Washington, DC.

U.S. Environmental Protection Agency (U.S. EPA), 2007. Water Security Initiative: Interim guidance on planning for contamination warning system deployment, EPA 817-R-07-002. EPA, Washington, DC.

U.S. Environmental Protection Agency (U.S. EPA), 2008a. Water Security Initiative: Cincinnati pilot post-implementation system status, EPA 817-R-08-004. EPA, Washington, DC.

U.S. Environmental Protection Agency (U.S. EPA), 2008b. Water Security Initiative: Interim guidance on developing an operational strategy for contamination warning systems, EPA 817-R-08-002. EPA, Washington, DC.

U.S. Environmental Protection Agency (U.S. EPA), 2008c. Water Security Initiative: Interim guidance on developing consequence management plans for drinking water utilities, EPA 817-R-08-001. EPA, Washington, DC.

U.S. Environmental Protection Agency (U.S. EPA), 2009a. CANARY user's manual and software upgrades. EPA/600/R-08/040A. EPA, Washington, DC.

U.S. Environmental Protection Agency (U.S. EPA), 2009b. Distribution system water quality monitoring: Sensor technology evaluation methodology and results, EPA 600/R-09/076. EPA, Washington, DC.

U.S. Environmental Protection Agency (U.S. EPA), 2010a. Detection of biological suspensions using online detectors in a drinking water distribution system simulator, EPA/600/R-10/005. EPA, Washington, DC.

U.S. Environmental Protection Agency (U.S. EPA), 2010b. Water quality event detection systems for drinking water contamination warning systems, EPA/600/R-10/036. EPA, Washington, DC.

U.S. Environmental Protection Agency (U.S. EPA), 2012a. Detection of contamination in drinking water using fluorescence and light absorption based online sensors, EPA/600/R-12/672. EPA, Washington, DC.

U.S. Environmental Protection Agency (U.S. EPA), 2012b. Threat Ensemble Vulnerability Assessment—Sensor Placement Optimization Tool (TEVA-SPOT) Graphical user interface user's manual, version 2.3.1, EPA/600/R-08/147. EPA, Washington, DC.

Ulitzur, S., Lahav, T., Ulitzur, N., 2002. A novel and sensitive test for rapid determination of water toxicity. Environ. Toxicol. 17 (3), 291–296.

van der Schalie, W.H., James, R.R., Gargan, T.P., 2006. Selection of a battery of rapid toxicity sensors for drinking water evaluation. Biosens. Bioelectron. 22 (1), 18–27.

van der Schalie, W.H., Shedd, T.R., Knechtges, P.L., Widder, M.W., 2001. Using higher organisms in biological early warning systems for real-time toxicity detection. Biosens. Bioelectron. 16 (7–8), 457–465.

van der Schalie, W.H., Shedd, T.R., Widder, M.W., Brennan, L.M., 2004. Response characteristics of an aquatic biomonitor used for rapid toxicity detection. J. Applied Toxicol. 24 (5), 387–394.

Wang, J., Lu, J.M., Hocevar, S.B., Ogorevc, B., 2001. Bismuth-coated screen-printed electrodes for stripping voltammetric measurements of trace lead. Electroanalysis 13 (1), 13–16.

Watson, J.P., Murray, R., Hart, W.E., 2009. Formulation and optimization of robust sensor placement problems for drinking water contamination warning systems. J. Infrastructure Sys. 15 (4), 330–339.

Weickgenannt, M., Kapelan, Z., Blokker, M., Savic, D.A., 2010. Risk-based sensor placement for contaminant detection in water distribution systems. J Water Resour. Plann. Manag. ASCE. 136 (6), 629–636.

Xie, C., Mace, J., Dinno, M.A., Li, Y.Q., Tang, W., Newton, R.J., Gemperline, P.J., 2005. Identification of single bacterial cells in aqueous solution using conflocal laser tweezers Raman spectroscopy. Anal. Chem. 77 (14), 4390–4397.

Xu, J.H., Johnson, M.P., Fischbeck, P.S., Small, M.J., VanBriesen, J.M., 2010. Robust placement of sensors in dynamic water distribution systems. Eur. J. Oper. Res. 202 (3), 707–716.

Yang, L.X., Chen, B.B., Luo, S.L., Li, J.X., Liu, R.H., Cai, Q.Y., 2010. Sensitive detection of polycyclic aromatic hydrocarbons using CdTe quantum dot-modified TiO_2 nanotube array through fluorescence resonance energy transfer. Environ. Sci. Technol. 44 (20), 7884–7889.

Yang, Y.J., Haught, R.C., Goodrich, J.A., 2009. Real-time contaminant detection and classification in a drinking water pipe using conventional water quality sensors: Techniques and experimental results. J. Environ. Manage. 90 (8), 2494–2506.

Zurita, J.L., Jos, A., Camean, A.M., Salguero, M., Lopez-Artiguez, M., Repetto, G., 2007. Ecotoxicological evaluation of sodium fluoroacetate on aquatic organisms and investigation of the effects on two fish cell lines. Chemosphere 67 (1), 1–12.

Response

1. INTRODUCTION

Threats must be evaluated and managed using systematic approaches and timely responses. As threats are progressively evaluated and confirmed, the responses become more significant. During threat evaluation, affected sites must be characterized and samples collected. Data collected from various sources are used to identify contaminants and confirm contamination events. As threats progress from possible to credible, analytical tests are performed on the samples collected and additional site data are collected. If a contamination event appears credible, operational and public health responses must be carried out. A challenging aspect of responding to water contamination threats is that the responses must be carried out quickly and appropriately.

2. THREAT EVALUATION

Threat warnings are issued following notification of a water contamination threat. These threats must be managed to determine if a water contamination incident is indeed occurring. It is important to identify water contamination events, but it is also important to avoid false alarms, so that the public's confidence is maintained. Also, responding to false alarms or not knowing whether or not a security breach is related to a contamination attempt can cause water utilities unnecessary cost in responding to perceived or potential threats (Moses and Bramwell, 2002). In characterizing water contamination threats, both the possibility and probability of these threats must be considered (U.S. EPA, 2003a). The source of a threat warning can be any of the following: signs of a security breach, witness account, notification by perpetrator, notification by law enforcement, notification by news media, unusual water quality measurements compared with baseline values, consumer complaint, or public health notification (U.S. EPA, 2002, 2003a). Threat warnings may also be issued by the Water Information Sharing and Analysis Center (WaterISAC). Following a threat warning, an evaluation of the threat must proceed, which consists of gathering information to determine if the threat is possible and credible, then if the threat can be confirmed by the results of analytical analyses.

The water contamination threat evaluation process has three stages. Initially, following issuance of a threat warning, threats are considered "possible." Once a threat warning has been issued, site characterization must be done to further confirm or reject the possibility of the threat. Data and samples are collected. Precautionary responses may also be undertaken to protect the public and the safety of response personnel. To minimize the spread of the contaminant and the exposure to the public, containment efforts might be made, if the source of the contamination event can be established. If the initial assessment suggests that the threat is possible, the threat is elevated to the next stage, that of being "credible." At this stage, steps are taken to confirm and identify the contaminant by analytical testing of samples collected. At this point, public notices may be issued advising the public not to use or drink the water. If the results of the testing analyses indicate that a contaminant is present, the threat is "confirmed." At this stage a crisis response is initiated, alternative water sources are identified, the portions of the affected water system are contained, and steps are taken to recover and rehabilitate the existing water system.

The threat evaluation process is continuous and iterative; the process results in repeated steps to identify contaminants, determine the extent of contamination, and initiate appropriate public health responses. Site characterization and sampling may not lead to successful identification of the contaminant(s), and repeated sampling and analysis may be necessary. In establishing the potential consequences of a contamination event, it will most likely be necessary to use system knowledge and data. A consequence analysis should be conducted that includes determining the number of individuals potentially affected, the severity of health effects, and potential effects of an interruption of the drinking water supply. A matrix that can be used in response planning is contained within the U.S. EPA Response Protocol Toolbox (Appendix 8.1 in U.S. EPA, 2003b). The matrix includes incidences and responses for varying threat levels. The responses include possible actions and the anticipated effects on the public (U.S. EPA, 2003b).

Establishing this type of matrix in advance is important in providing guidelines for responders during threat evaluation.

To formulate a response, water providers must rely on the data they can collect in a short time period regarding:

- Physical evidence.
- Credibility of threat information and eyewitness reports.
- Location-specific information.
- Results of on-site water quality testing.
- Background concentrations of the contaminant at the site (if applicable).
- Tentative results from laboratory analyses.
- Information about the contaminant(s) of interest, including toxicity and environmental fate.
- Public health information, including evidence of illness or infection in the community (e.g., deaths, emergency room visits, sale of over-the-counter medications).

An evaluation of consequences follows from this list of information that must be collected. Multiple sources of information must be collected and evaluated to develop a response. The training documents described in Chapter 2 contain forms that water utilities can fill out during a threat evaluation, and software packages are available as well. Also, a worksheet for assisting water providers during threat evaluation is contained in the U.S. EPA Response Protocol Toolbox (Appendix 8.2 in U.S. EPA 2003b).

Water providers must aggressively plan ahead of time to be prepared for water contamination events. A complete and detailed emergency response plan (ERP) must be developed ahead of time. The U.S. EPA (2003b) recommends a target time of 1 hour for determining if a threat is possible and a target time of 2–8 hours for determining if a threat is credible. These time frames do not allow sufficient time to formulate a response strategy and work plan. In a review of six outbreaks that appeared to be the result of intentional contamination, it was determined that it took 0–6 days to notify the CDC of the outbreak and 0–26 days to identify causative agent (Ashford et al., 2003). Plans and procedures must be developed ahead of time to facilitate quick responses. Then, during threat evaluation, water providers should be consulting their ERPs for guidance.

An outline of the threat management process is contained in Table 6.1. It is important that water utilities and first responders proceed with these steps quickly in responding to a water threat.

2.1. Communication Strategy

During threat evaluation, a communication strategy must be in place, which should be established prior to the threat, during the planning stages (U.S. EPA, 2012a). The chain of command should be documented in the

TABLE 6.1 Threat Management Process (U.S. EPA, 2004, Figure 5-1)

Steps	Appropriate Actions
Determine public health consequences due to water contamination	Evaluate contaminant properties Assess spread of contaminant
Implement operational responses	Isolate and contain potential contaminated water, if possible Consider novel operational responses
Implement public notification strategy	Provide public notification
Implement alternate water supply	Provide appropriate alternate water supply
Return to normal operation and use	Notify the public Demobilize alternate water supply

ERP and put into action during threat evaluation. Also, a notification hierarchy should already be established, with the entire operation being headed by the incident commander (IC). The following agencies and organizations may need to be contacted during threat evaluation: Federal Bureau of Investigation, Centers for Disease Control (CDC), U.S. EPA National Response Center, neighboring utilities, wastewater utilities, U.S. EPA regional offices, local emergency responders (fire, emergency medical services, hazmat), state emergency responders, local government officials, state government officials, state environmental and public health agencies, local health departments, public health and environmental laboratories, state environmental departments, local law enforcement, state law enforcement, and the media (U.S. EPA, 2007b). The contact information for all potentially involved parties should be contained within the ERP documents. The IC remains in place during the recovery and rehabilitation stages of a water contamination event; however, the structure and involved parties are different during these phases compared with the structure in place during the immediate response.

Effective communication is critical for an effective response to a water system threat (Jalba et al., 2010). In reviewing public records of water contamination events in the United Kingdom, Gray (2008) found that closer coordination is needed between water utilities and public health officials and that there were often delays in obtaining analytical results and health effects information.

Communication with the public can occur using a utility hotline with a recorded message, a utility call center, a utility website, door-to-door notification, reverse phone calls, prescribed messages, press releases, and public service announcements (U.S. EPA, 2008). Message mapping is a communication tool that allows water utilities to respond quickly and efficiently by preparing communication during the planning stages (U.S. EPA, 2007a).

2.2. Information Management

To evaluate and define threats, information needs to be collected and reviewed. To assist in this process, use of crisis information management software (CIMS) and the field operations and records management system (FORMS) is recommended (U.S. EPA, 2003a). Several commercial CIMS packages are available, and this software is used by emergency management agencies during emergency events to record information. Depending on the CIMS package, the software can be integrated with GIS software. CIMS packages are compliant with an incident command system (ICS), making it easier to integrate these programs into emergency response efforts. The software can also be used as part of planning. The FORMS software was developed by the U.S. EPA Contract Laboratory Program. FORMS is used to manage records for sample documentation, analysis, and tracking, which can be useful during a water threat response effort.

The main tasks included in information management are site characterization, sample collection, sample analysis, and collection of public health information. It may be necessary to collect information from multiple sources, including state regulatory agencies, the U.S. EPA, law enforcement agencies, the FBI, neighboring utilities, public health agencies, 911 call centers, the Water Information Sharing and Analysis Center (ISAC), the Department of Homeland Security warning and alerts, and contaminant information from the CDC (U.S. EPA, 2003b). The WaterISAC is a national resource that serves as a clearinghouse for alerts, warnings, and information. Consultation of the WaterISAC during a threat evaluation is recommended, as this organization can assist in locating information regarding contaminants and laboratory resources. Collecting data on previous threats and security incidents can help in the assessment of the validity of the current threat.

For a threat to be elevated to the confirmatory stage, analytical confirmation of the contaminant is necessary. At this point, it is important to collect sufficient information and data to confirm the results and interpret the data collected. Strict and standardized quality assurance/quality control (QA/QC) protocols must be followed to validate the results. Additional site characterization and public health confirmation should also be completed to corroborate the analytical results. Once a threat has been confirmed, recovery and rehabilitation activities should commence, including additional characterization and feasibility studies.

3. POTENTIAL RESPONSES

Water providers may use numerous responses once a contamination event is detected with some certainty (Table 6.2). The responses employed depend on the degree of certainty in the declaration of the contamination event, the facility or component attacked, and the potential effect (U.S. EPA, 2002, 2006). If water contamination is possible, water providers are likely to proceed with

TABLE 6.2 Potential Responses to a Water Contamination Threats

Potential Response	Examples
Increase monitoring	Site evaluations Test water using rapid on-site test kits Collect samples for additional laboratory analyses Send samples to a specialized laboratory Monitor water pressure, temperature, etc.
Increase surveillance	Evaluate customer complaints Collect public health data
Increase security	Notify employees Increase oversight of operations Deploy guards at essential facilities
Public announcements	Boil advisory Do not drink advisory Do not use advisory
Shut off system or part of system	Operate isolation valves Discontinue use of wells, pump stations, etc. Reduce system pressure
Additional public warnings	Reduce system pressure Add dye to alert consumers
Extract contaminated water	Use hydrants to extract water Drain reservoirs
Provide alternate water supply	Distribute water bottles Distribute water using tank trucks Provide alternative source with same distribution system
Change treatment	Increase chlorine dosage Additional booster chlorination Mobilize treatment units throughout system Reduce filtration rates

at least providing increased monitoring, surveillance, or security. Increased diligence in operating the water system could take many forms. Water providers may choose to take more samples for water quality analyses or perform additional tests on the samples collected. Samples should be sent to appropriate outside analytical laboratories if warranted. It may be possible to allow contaminants to be reduced through natural attenuation or degradation. The water system most likely would be monitored intensely as part of this no-action alternative.

3.1. Site Evaluation

During threat evaluation, the site(s) of suspected contamination should be evaluated to better define the threat. Signs of a security breach may include unauthorized intrusion into a secured facility, which may be based on an alarm, cut lock, open door, cut fence, or other anomaly. All security breaches must be evaluated to determine if deliberate contamination occurred, since security breaches may also be the result of vandalism or theft. Site responders should look for signs of contaminants, such as empty containers or pumps. Responders need to define the boundaries of the affected site and perform an initial investigation that includes a thorough observation of site conditions.

The response team should also collect samples and document their observations. The U.S. EPA Response Protocol Toolbox (Appendices 8.3–8.6 in U.S. EPA, 2003b) contains the following forms to assist in threat evaluation: Security Incident Report Form, Witness Account Report Form, Phone Threat Report Form, and Written Threat Report Form. These forms should be used to allow standardized and uniform threat reporting. Notifications should be made to WaterISAC and other agencies. The requirements for threat verification and reporting should be established in the ERP. As threat evaluation progresses, site evaluation continues and may affect the progression of the threat status. Site evaluation affects operational and public health responses, resulting in an iterative process to characterize and respond to potential contamination.

Responses to water threats need to proceed quickly. Delays in reporting biological terrorism events significantly affect the response effort (Ashford et al., 2003). A swift response to natural disasters and other events is necessary to maintain public health and societal order (Watson et al., 2007).

3.1.1. Site Safety

An important aspect of site characterization is to evaluate the site to determine if it is safe for responders to be present. Ensuring the safety of the responders may require mobilization of hazmat teams, the FBI, and other agencies in addition to responders from the water utility that are familiar with the water system. Trained personnel are needed for site evaluation and personal protective equipment (PPE) is needed (CDC, 2011). The use of robots to collect information and samples has even been suggested. The CDC and WHO provide guidance on safety and appropriate response procedures (U.S. EPA, 2004; WHO, 2002, 2004). A health and safety plan should be developed as part of the planning process, and this document should be consulted during emergency response site evaluations. The PPE should be available prior to threats, and the response team should be trained in its use. During the hazard assessment of the site, testing for radiological isotopes and toxic chemicals should be conducted using handheld devices. The presence of radioactive or volatile chemicals indicate an immediate threat, requiring continued use of PPE and trained hazard response personnel.

3.1.2. Rapid On-Site Testing

As part of the initial investigation, it is recommended that rapid testing be completed. The U.S. EPA Environmental Technology Verification program maintains information on handheld monitoring and detection technologies that can be used in responding to a water contamination threat. In addition, the Association of Analytical Communities maintains a database of rapid test kits with kits available for bacteria (general), *Legionella, Bacillus, Campylobacter*, coliforms, *E. coli* (including O157 and H7 STEC), *Listeria, Salmonella, Salmonella*, and *Staphlococcus*. Rapid field testing of samples can assist in tentative identification of contaminants, which assists in determining if special handling of the collected samples is needed. Rapid testing has been shown to be effective, although sensitivity and specificity are of concern; kits are available for testing pathogens, biotoxins, nerve agents, pesticides, toxicity, and a variety of other chemicals (Dewhurst et al., 2002; States et al., 2004; van der Schalie et al., 2004). Field deployable PCR and GC/MS units are available (States et al., 2004). Testing for standard water quality parameters also is helpful for determining if any of these constituents are deviating from typical values. Both core and expanded field test kits should be utilized (Magnuson et al., 2005; U.S. EPA, 2003c, Table 3-3). Core testing should include radioactivity, cyanide, chlorine residual, pH, and conductivity, while expanded testing should include general hazards, volatile chemicals, Schedule 1 chemical weapons, water quality parameters, pesticides, volatile organic compounds, biotoxins, pathogens, and toxicity (Magnuson et al., 2005). As noted by Hrudey and Rizak (2004), the occurrence of false-positive and false-negative results must be considered when interpreting the results of all analyses, especially when evaluating results from rapid testing. The U.S. EPA Response Protocol Toolbox (U.S. EPA, 2003c) contains forms that are useful in site characterization, including a site characterization plan template, a site characterization report form, a field testing results form, a sample documentation form, and a chain of custody form. Use of these forms assist in collecting consistent and complete records of any incident.

In addition to collecting samples, potentially contaminated water should be tested using rapid field test kits to provide a safety screening (States et al., 2004, 2006). If unsafe site conditions have been detected, hazmat protocols may need to be followed, not only for site access but also for sample collection. If the samples are considered hazardous, special shipping requirements may be needed. Also, measurements of standard water quality parameters (e.g., pH, conductivity) can be made to further assess if any of these parameters are different than normally observed values. For standard water quality parameters to be useful, it is necessary to establish baseline water quality information with which to compare data collected during a threat.

3.1.3. Consumer Complaint Data

To further define the affected site and facilities, information should be collected from consumers. One way to better understand the extent of contamination is through consumer complaints. The U.S. Army Center for Health Promotion and Preventative Medicine (2003) has a guide on investigating consumer complaints that can be helpful in interpreting the data collected. Whelton et al. (2004) describes the guide for interpreting consumer complaints and using this information to detect a water system failure or contamination. Also, the U.S. EPA Response Protocol Toolbox (U.S. EPA, 2003b) has a Water Quality and Consumer Complaints Report Form and a Public Health Information Report Form. Customer complaint surveillance data can be recorded and tracked electronically, so that it is compared with other types of data collected to alert of a threat (Allgeier et al., 2011).

Consumer complaints can arise from changes in water quality that are not related to security events, so it is important to distinguish between these causes in responding to a threat. The types of complaints that may be received by water utilities fall under the following categories: aesthetics (color, taste, and odor), water pressure, illness, and suspicious activity (Whelton et al., 2004). The taste and odor characteristics for several important chemical contaminants, such as cyanide, are described by Whelton et al. (2004) as well as a presentation of case studies where consumer complaints led to detection of a water system problem. In some cases, consumer complaint information has limited usefulness. For example, it is unlikely that a biological contaminant would greatly affect the taste and odor of drinking water (Nuzzo, 2006).

3.1.4. Public Health Data

In addition to using consumer complaints to define the boundaries of contamination, public health data can also be used to better understand the extent of the contamination. These data are available from hospitals and public health officials. Public health surveillance data can include reportable diseases reporting and syndromic reporting (Roberson and Morley, 2005). Disease reporting occurs through medical records of certain illnesses and symptoms and death records. Examples of syndromic public health reporting include sales of over-the-counter medications, emergency room visits, orders for laboratory tests, poison control center calls, and other measures of illness (U.S. EPA, 2007b).

3.1.5. Water System Data

To estimate the contaminated area, information regarding the suspected location and estimated time of contamination is needed. To determine the spread of the contaminant, supervisory control and data acquisition (SCADA) systems and operator knowledge are used. Use of hydraulic models (e.g., EPANET, PipelineNet, MWHSoft, Stoner, Haestad) may also be helpful, although the timely use of these models may not be possible for determining the extent of the contaminated portions of the water system. Water system modeling simulations should be used as planning tools to understand contaminant migration through the water distribution system (U.S. EPA, 2003b) and most likely will be useful in the rehabilitation stage to assist in recovery efforts. Confirming a threat may be a time-consuming process, and a response should not be delayed by data collection and contaminant characterization activities. The RPTB contains an outline of the steps that should be taken to determine the credibility of threats and appropriate responses (U.S. EPA, 2003b).

3.2. Sample Collection

Sampling is an important part of threat evaluation, although many of the definitive tests that could be used to confirm a threat are time consuming and require transport of samples to specialized laboratories. Since water contamination has the potential to cause loss of life and damage to property, it is important to collect different types of data on potentially contaminated water samples. Samples should be collected for multiple chemical and biological analyses using standardized sample collection techniques. To assist in collecting samples correctly, predesigned sample kits are available (U.S. EPA, 2003c, see Tables 3-1 and 3-2 for examples). Samples may require pretreatment, such as chemical quenching and acidification, at the time of sample collection. In tests for biological analyses, especially protozoan and virus detection, large volumes of water may need to be filtered so that the retained solids can be eluted and assayed. According to the U.S. EPA (2003c), 4 liters of water is needed for bacterial analysis and 10 liters for protozoan analysis. Ultrafiltration is a proven method for concentrating samples for analysis of contaminants present in small concentrations; however, membrane filtration has also been shown to be effective for rapid field testing (States et al., 2006). Larger sample sizes may be necessary if the concentration of pathogenic contaminants is low. Linquist et al. (2007) used ultrafiltration to concentrate and detect *Bacillus anthracis*, *Bacillus atrophaeus* subspecies *globigii*, and *Cryptosporidium parvum* from 100 liter samples. Protocols for filtration should be established prior to a contamination event and should be part of the standard operating procedures (SOPs) written in advance and tested. During sample collection, care should be taken to avoid production of aerosols, as hazards can be spread through aerosols. Dilution of samples may make them less hazardous for transport and analysis, although it is typically not recommended because of the effect dilution has on detection limits. Reduced sample volumes may reduce exposure of responders to hazards, although it may be difficult to adequately analyze samples with completely unknown composition if large samples volumes are not available.

Guidance is provided on where to direct sampling efforts during a water threat. Computational methods can be used to determine the most effective locations for expanded sampling (Eliades and Polycarpou, 2012; Eliades et al., 2011). These methods must be in place prior to emergency situations, and water utility operators must be proficient in using them.

To definitively determine the presence of a water contaminant, rigorous QA/QC procedures need to be followed. Responders should clearly document their findings, including all measurements, in field data sheets and logbooks. All equipment should be calibrated and checked before and after deployment. Specific QA/QC procedures should be determined ahead of time and documented in SOPs, so that they are established at the time of a threat assessment. The SOPs for sampling and analysis should be contained within the ERP document made available to threat responders. Since many conventional laboratory analyses are time consuming, it may be necessary to provide a tentative analysis and revise the results as subsequent QA/QC tasks are completed.

Once an initial evaluation has been made regarding the type of contaminant present, sampling can be directed toward a biological hazard, chemical hazard, or radiological hazard (U.S. EPA, 2003c, see Figure 3-3). Correct initial characterization of samples allows for reduced testing, reduced sampling, and collection of appropriate sample volumes. The SOPs should be organized to assist selection of appropriate methodologies in response to the initial assessment.

In contamination and outbreaks, it is often difficult to attribute the cause to a specific agent, particularly a biological agent (Christopher et al., 1997). By the time that an event is noted, the water may no longer be contaminated or the contaminant concentrations may be too low to permit verification.

3.2.1. Laboratory Resources

Once a threat is elevated from possible to credible, the samples collected during the possible stage require analysis. Since the nature of the contaminant(s) is likely unknown, special handling and analyses are required. Most water utilities lack the capability to perform many of the analytical tests. Water utilities are unlikely to have the equipment, trained personnel or the laboratory certifications to perform many of the tests needed to detect hazardous chemical and pathogenic microorganisms. Their labs are typically set up to perform environmental testing and routine microbial testing of indicator organisms. To analyze samples for more complex and dangerous chemical and biological constituents, it is likely that they would have to outsource the testing to more qualified laboratories.

A nationwide network of laboratories is titled the Integrated Consortium of Laboratory Networks. Within this network is the Environmental Response Laboratory Network (ERLN), an "all hazards" and environmental laboratory network, and it includes laboratories that can detect toxic industrial chemicals and chemical warfare agents, biological agents, and radiochemical agents (Antley and Mapp, 2010). The ERLN provides an *Environmental Laboratory Compendium*, which is a database of nationwide environmental laboratories that contains information about analytical capabilities. Registration is required to access the database. The ERLN is run by the U.S. EPA and operates under the ICS structure, making it easier for water utilities to communicate during crisis situations. Within the ERLN is the Water Laboratory Alliance (WLA), an integrated nationwide network of laboratories that is prepared to handle samples collected during contamination events (U.S. EPA, 2012b). The Centers for Disease Control (CDC) developed the WLA to specifically address analysis of biological and chemical weapons. In addition, the WLA provides training to representatives from water utilities. In the event of confirmed water contamination, sampling should continue to gauge the progress of recovery and rehabilitation efforts.

A laboratory network parallel to the ERLN is the Laboratory Response Network (LRN), which is a network of highly specialized laboratories developed by the CDC, the Association of Public Health Laboratories, and the FBI for dealing with bioterrorism threats. The laboratories in the LRN can analyze many pathogens and some biotoxins, but they are usually not qualified to test for chemical weapons. While a few laboratories can work with dilute chemical weapons, only two laboratories in the United States are capable of working with concentrated Schedule 1 chemical weapons: the U.S. Army Edgewood Laboratory and Lawrence Livermore National Laboratory (U.S. EPA, 2003d). Laboratories that analyze environmental samples and perform routine chemical and biological analyses most likely are not qualified to handle and assay samples from a deliberate attack. For samples potentially containing chemical weapons and biotoxins, it is necessary to use specialty laboratories and those associated with the LRN.

During a water contamination event it may also be necessary to collect and test clinical samples to confirm the causes of illness and possibly death. The clinical and environmental samples most likely need to be handled separately. In addition, the Food Emergency Response Network (FERN) may be helpful in a laboratory response to water contamination. For radiological contaminants, the Federal Radiological Management Center (FRMAC), which is operated by the Federal Emergency Management Agency (FEMA) may be helpful.

In addition, many states have laboratory networks of their own and provide assistance to water utilities in sampling for suspected contaminants (Crisologo, 2008). For example, the California Mutual Aid Laboratory Network (CAMAL Net) is modeled after the Water/Wastewater Agency Response Network (WARN). The CAMAL Net is a network of California Department of Public Health (CADPH) labs, the U.S. EPA, California state agencies, and public water utilities. In addition, the California Health Alert Network (CAHAN) collaborates during public health emergencies.

The U.S. EPA RPTB has examples of how laboratories respond to water threats, including how existing utility and commercial laboratory resources are utilized (U.S. EPA, 2003d, Sections 7.1 and 7.2).

3.3. Sample Analysis

A challenge in analyzing samples collected from potentially contaminated sites is that typically the nature of the contaminants is unknown. As such, screening protocols for unknown chemicals should be followed. Basic screen procedures utilizing established analytical techniques and standard methods should be utilized. Standard methods should be used to the extent possible, although it may be necessary to use nonstandard methods for constituents not typically found in drinking water samples. In addition, an expanded screening should be done utilizing exploratory techniques that may not have standard methods (e.g., immunoassays, LC/MS). Establishing the chain of custody during sampling and testing is needed to provide scientific and legal evidence for the analysis. Samples should be stored at proper temperatures (e.g., 4°C) until analyses are completed. In all testing procedures, the possibility of false positives and false negatives should be considered.

3.4. Collection of Contaminant Information

Once a preliminary confirmation of the type of contaminant is determined, information should be collected that can be used in the response effort. Information on various contaminants is available through various sources. The U.S EPA Water Contaminant Information Tool (WCIT) is a secure database that contains information on potential contaminants, including treatability information. The Water ISAC also provides information on various contaminants. Laboratories within the Water Laboratory Alliance and Environmental Response Laboratory Network also have information on potential contaminants. Many of the contaminants that could be present as the result of accidental or deliberate contamination are typically not found in drinking water, and utilities may not have familiarity with these chemicals, including toxicity, appropriate analytical procedures, and the like.

Information on the contaminant helps guide the public health and operational responses. Bryant and Abkowitz (2007) present work for a response management system for chemical spills on land. Guidance provided in their work can also be useful in the context of water security. For example, Bryant and Abkowitz (2007) recommend determining the toxicity, reactivity, flammability, persistence, vapor pressure, solubility, dynamic viscosity, specific gravity, conductance, and adsorption isotherm parameters for potential chemical contaminants.

3.5. Operational Responses

Once a threat is identified as being possible, it is necessary to immediately develop appropriate operational responses to protect public health (U.S. EPA, 2002, 2003b). Possible operational responses should be identified in planning documents, so that they can be acted on quickly. One of the most likely operational responses involves containment to reduce the spread of contaminants and reduce the likelihood of exposure to the population served by the water system (Eliades and Polycarpou, 2012). To adequately contain affected portions of a water system, it is necessary to estimate the contaminated area, which may necessitate estimating the location of contaminant introduction and the spread of the contaminant. Information collected on water quality, consumer complaints, and public health data are useful in defining the areas affected. Extracting contaminated water from the area near the contamination site prevents the spread of contamination by forcing water to flow to the source of contamination instead of away from the source. The potential effects on consumers and firefighting need to be considered. Individual responses depend on the characteristics of the water system and the nature of the contaminant. Operational responses are modified as more information is collected on the contamination.

Knowledge of a contamination event may also lead water providers to alter their treatment and disinfection strategies. Opening and closing of valves can be guided using sensor network software to minimize the contamination impact (Guidorzi et al., 2009; Poulin et al., 2008). Preis and Ostfeld (2008) developed a contaminant response model that took into account both public consumption of the contaminant and the number of actions (opening and closing valves) needed to isolate the contaminant. Public consumption of the contaminant was given by:

$$F_1 = \sum_{i=1}^{N} \sum_{t=t_d}^{EPS} C_i(t) \times V_i(t) \tag{6.1}$$

where F_1 is the mass of contaminant consumed after detection occurs, N is the total amount of public nodes, EPS is the extended period simulation or total simulation run time, i is the node index, t is the time elapsed after the initial detection time t_d, $C_i(t)$ and $V_i(t)$ are the contaminant concentration and volume of consumed water at node i at time t, respectively (Preis and Ostfeld, 2008). The number of valve closures and hydrant openings was optimized using the equation:

$$F_2 = \sum_{k=1}^{VA} VA_k + \sum_{j=1}^{HY} HY_j \tag{6.2}$$

where VA and HY are the total number of valves and hydrants, respectively; k and j are the valve and hydrant indexes, respectively; and VA_k and HY_j are set to 1 if the kth valve was closed or jth hydrant was opened, respectively, but are otherwise zero (Preis and Ostfeld, 2008). Multiobjective optimization was done by optimizing:

$$F(x) = [f_1(x), f_2(x), ..., f_M(x)]^T \tag{6.3}$$

under the conditions:

$$g_i(x) > 0, i = 1, 2, ..., k$$
$$e_j(x) = 0, j = 1, 2, ..., l \tag{6.4}$$

where $x = (x_1, x_2, \ldots, x_n)^T$ are the vector of decision variables; k is the inequality constraint; and l is the equality constraint (Preis and Ostfeld, 2008).

Mobilizing booster chlorination units to locations throughout the distribution system can be an effective disinfection and treatment strategy. Betanzo et al. (2008) showed, using water system modeling software, that maintaining standard chlorine residual concentrations throughout the distribution system is protective against many pathogens, although the efficacy depends on the type of pathogen. Guidance on locating booster chlorination stations throughout a distribution system is provided by Lansey et al. (2007). Another example is use of additional treatment stages at treatment plants. Baranowski and LeBoeuf (2006) used a sensor network approach with optimization calculations to simulate contamination events in modeled networks, where detection of an event is followed by operational responses (draining of the contaminated water) accomplished in the model using reduced demand at nodes. The results of Baranowski and LeBoeuf (2006) demonstrate that sensor network resources can be adapted to be used in responding to threats.

3.5.1. Impact of Operational Responses on Firefighting

One risk associated with discontinuing water service is that firefighting capability is compromised. Those responsible for contamination events should be aware that contaminating the water supply may also affect firefighting, and it is important that communities be prepared to respond to simultaneous attacks. It is possible to maintain the water supply and minimize the risk of exposure to consumers; however, the most appropriate response depends on the type of contaminant (information that may not be known during the attack). Water system pressure can be maintained at a minimal level (e.g., 20 psi) to provide some firefighting capability. The volume of water needed for fighting fires can be reduced through the addition of certain chemical compounds. In addition, it may be possible for water providers to work closely with firefighting agencies during an emergency to determine if the public water service must be temporarily restored. It would also be possible to provide an alternative firefighting supply so that the system water pressure can be reduced.

3.5.2. Feasibility of Real-Time Responses

The capability to provide a real-time response to isolate and potentially stabilize contaminants is desired; however, the human element in decision-making cannot be avoided. A real-time response involves detecting a contamination event using in-line detection sensors and analysis software, then allowing the system to automatically conduct a response, such as isolating portions of the system or shutting off the water. Real-time operation of valves can improve water quality and optimize operation of booster chlorination within a distribution system; real-time operation of valves could also be used to respond to water contamination (Kang and Lansey, 2010). The possibility of a real-time response is an elusive goal at this point in time. A real-time response requires a lot of confidence in the contaminant detection capabilities. A pilot contaminant warning system was implemented for the city of Cincinnati, and one of the main problems encountered was false alarms (Allgeier et al., 2011; Fencil and Hartman, 2009). It seems that, at this point in time, considerable human judgment needs to be involved in determining if a threat is credible. However, responses need to be timely. Ashford et al. (2003) report that it has taken as long as 26 days in the past to determine the cause of an outbreak.

A real-time response could take many forms. Kang and Lansey (2010) consider real-time operation of valves and use of booster disinfection using computer simulations to optimally apply chlorine throughout a distribution system. Parks and VanBriesen (2009) modeled the use of booster chlorination for real-time response to a biological contaminant and found that the effectiveness of booster chlorination depends on the location of chlorination injection and the site of contamination. Booster disinfection can be used to provide injection of chlorine at various points within a distribution system, which is beneficial for reducing total chlorine dosage across the entire system while maintaining minimum chlorine residual (Betanzo et al., 2008; Parks and VanBriesen, 2009). Use of booster chlorination as a response tool would be applicable only if the contaminant is responsive to inactivation or removal by chlorination. Another potential real-time operational response is flushing. Haxton and Uber (2010) investigated the use of flushing following real-time detection of a contaminant and found that the effectiveness is related to the ability to detect the source, location, and general distribution of the contaminant.

3.6. Public Health Response

As defined by the U.S. EPA (2003e), public health responses are "actions taken to mitigate consequences resulting from threats or incidents involving biological, chemical, or radiological contaminants" (p. 16). The public health consequences of water contamination events need to be determined to support health response decisions, although it may be difficult to do so. At the credible stage, public health responses include "do not drink" notices and "do not use" notices. Depending on the nature of the contaminant, water can be provided while consumers are warned not to consume it or use it for other purposes such as bathing. At the confirmatory stage, public health responses should be revised as remediation and recovery plans are developed. At this stage, a communications and public relations plan should be put into place (CDC, 2013). Interacting with public health officials is an important aspect of water security planning and response to threats. Water utilities need to communicate their activities to the public health agencies, and it is possible that a water contamination threat notice will originate with the public health agencies (U.S. EPA, 2002).

Implementation of a public notification strategy should include public announcements and instructions. Communication with the public is important for retaining social order and public confidence. Public announcements are one way to continue operating the water system, while minimizing exposure to potentially contaminated water. Guidance is available for communicating with the public and with the media during an emergency drinking water threat, including standard notifications and message mapping (CA DHS, 2006). The disadvantage of public announcements is that not all consumers may be reached. Methods used to transmit public announcement include reverse 911 calls, bullhorn announcements (from ground or air), door-to-door visits, radio announcements, email, text messages, and newspaper advertisements. Injection of dye is one way to warn consumers, although water providers would need to have access to the dyes ahead of time and a plan to disperse the dye. The public should be aware of the significance of the dye. It would be possible to withdraw water near the source and inject the dye in the perimeter of the area to disperse the dye.

Public health responses used in a contamination event must be appropriate for the specific contaminant. For example, advising residents to boil water prior to consuming it is a good response for eliminating exposure to many pathogens. However, some pathogens, including *Bacillus anthracis*, are resistant to heat and can survive boiling (Rice et al., 2004). Experiments performed using a nonpathogenic strain of *Bacillus* demonstrated that the spores are spread via the steam as the result of boiling and can result in exposure through inhalation. Rice et al. (2004) found that boiling water in a covered container for 3–5 minutes resulted in at least four log removal *Bacillus* spores; less efficient disinfection occurred when the water was boiled in an open container. Additionally, bacteria that are known to grow in hot water systems, such as *Legionella*, could be transmitted through showering. Transmitting "boil" advisories or "do not drink the water" advisories may not be completely protective of public health.

Public health notices are not always effective. For example, during an outbreak of salmonellosis resulting from a contaminated public water supply, 31% of households did not follow the public health "boil water order" advisory and many of these resident became ill as a result (Angulo et al., 1997). In the outbreak investigated by Angulo et al. (1997) The main reasons that residents did not follow the "boil water order" were not remembering and not believing the order.

4. SHORT-TERM EMERGENCY WATER SUPPLY

Short-term alternative water supplies are needed in the event of a credible threat and long-term alternative water supplies are needed in the event of a confirmed threat. Guidance on procuring and distributing emergency water supplies is available, but it is also something that water utilities should consider in their planning efforts (U.S. EPA, 2011; WHO, 2002, 2004). Providing an alternative safe water source is imperative in responding to a disaster situation to maintain social stability (Watson et al., 2007).

Alternative water supplies can be provided using bottled water, water delivered in tank trucks, and treatment of water from alternative sources. Water may be available from neighboring communities. Water utilities can facilitate better responses to emergencies by identifying bottle water companies and tanker services, potentially establishing agreements ahead of time (U.S. EPA, 2008). Pre-established mutual aid agreements with other agencies may also be beneficial. The costs of alternative sources of water and sources of funding must be identified. It is possible that some water utilities may have backup groundwater supplies that can be accessed during an emergency. Procurement and distribution may be carried out using centralized or distributed organizational structures.

In planning for an emergency water supply, both the quantity and quality needed should be considered (U.S. EPA, 2011; WHO, 2002). Separate sources may need to be identified for firefighting. The roles and responsibility of those in charge of the emergency water supply must be established in accordance with the overall IC and should be in accordance with the National Response Framework. The security of the emergency water supply should be ensured using appropriate measures. The U.S. EPA (2011) recommends screening emergency water supplies based on the following criteria: reliability, ease of implementation, capacity, cost, transportation and distribution ease, and mobilization time.

Providing water supplies in the aftermath of a credible water threat most likely will be in the form of a temporary, short-term solution (e.g., providing bottled water for basic needs and delivering by truck to neighborhoods). If the water supply continues to be compromised during a lengthy recovery and rehabilitation effort, a more permanent long-term alternative supply must be found.

CONCLUSIONS

Threats must be evaluated and managed using systematic approaches and timely responses. As threat warnings are issued, progressively evaluated, and confirmed, responses become more significant. During threat evaluation, affected sites must be characterized and samples collected to identify contaminants and confirm a contamination event. Water providers must have emergency response plans in place so threat evaluation, confirmation, and response can proceed without delay.

Emergency responses to contamination threats include increasing sampling frequency, on-site testing, safety evaluations, monitoring public health and water system data, and isolating and extracting contaminated water. In addition, timely public health responses, such as warnings against water consumption, providing short-term water supplies, and public communication, are essential to protecting the general public. Public communication is especially important, as some people may not believe or follow recommendations from the water suppliers. Providing short-term water supplies, such as bottled water or tank trunks, can help ensure adherence to consumption advisories and maintain social stability.

A challenging aspect of responding to water contamination threats is that responses must be carried out quickly and appropriately. The technology for real-time response systems that require no human element, such as sensor networks that control system valves, requires further development but can ideally reduce response times and better protect public health. Other technologies, such as software that facilitates communication during emergencies, can also prove to be beneficial. As a result, opportunities exist for further research to develop rapid contaminant screening and real-time response systems, as they are necessary to provide accurate information and advisories to the general public and develop appropriate operational responses.

REFERENCES

Allgeier, S.C., Haas, A.J., Pickard, B.C., 2011. Optimizing alert occurrence in the Cincinnati contamination warning system. J. Am. Water Works Assoc. 103 (10), 55–66.

Angulo, F.J., Tippen, S., Sharp, D.J., Payne, B.J., Collier, C., Hill, J.E., Barrett, T.J., Clark, R.M., Geldreich, E.E., Donnell, H.D., Swerdlow, D.L., 1997. A community waterborne outbreak of salmonellosis and the effectiveness of a boil water order. Am. J. Public Health 87 (4), 580–584.

Antley, A., Mapp, L., 2010. ERLN/WLA launch. 2010 Water Laboratory Alliance (WLA) Security Summit, San Francisco, CA.

Ashford, D.A., Kaiser, R.M., Bales, M.E., Shutt, K., Patrawalla, A., McShan, A., Tappero, J.W., Perkins, B.A., Dannenberg, A.L., 2003. Planning against biological terrorism: Lessons from outbreak investigations. Emerg. Infect. Dis. 9 (5), 515–519.

Baranowski, T.M., LeBoeuf, E.J., 2006. Consequence management detection optimization for contaminant and isolation. J. Water Res. Plann. Manag. ASCE 132 (4), 274–282.

Betanzo, E.W., Hofmann, R., Hu, Z., Baribeau, H., Alam, Z., 2008. Modeling the impact of microbial intrusion on secondary disinfection in a drinking water distribution system. J. Environ. Eng. ASCE 134 (4), 231–237.

Bryant, D.L., Abkowitz, M.D., 2007. Development of a terrestrial chemical spill management system. J. Hazard. Mater. 147 (1–2), 78–90.

California Department of Health Services (CA DHS), 2006. Crisis and emergency risk communication workbook for community water systems. CA DHS. Sacramento, CA.

Centers for Disease Control and Prevention (CDC), 2011. Emergency response resources: Personal protective equipment. CDC, Atlanta, GA.

Centers for Disease Control and Prevention (CDC), 2013. Drinking Water Advisory Communication Toolbox, Public Health Service, U.S. Department of Health and Human Services, American Water Works Association. CDC, Atlanta, GA.

Christopher, G.W., Cieslak, T.J., Pavlin, J.A., Eitzen, E.M., 1997. Biological warfar, a historical perspective. JAMA-J. Am. Med. Assoc. 278 (5), 412–417.

Crisologo, J., 2008. California implements water security and emergency preparedness, response, and recovery initiatives. J. Am. Water Works Assoc. 100 (7), 30–34.

Dewhurst, R.E., Wheeler, J.R., Chummun, K.S., Mather, J.D., Callaghan, A., Crane, M., 2002. The comparison of rapid bioassays for the assessment of urban groundwater quality. Chemosphere 47 (5), 547–554.

Eliades, D.G., Polycarpou, M.M., 2012. Water contamination impact evaluation and source-area isolation using decision trees. J. Water Res. Plann. Manag. ASCE 138 (5), 562–570.

Eliades, D.G., Polycarpou, M.M., Charalambous, B., 2011. A security-oriented manual quality sampling methodology for water systems. Water Res. Manag. 25 (4), 1219–1228.

Fencil, J., Hartman, D., 2009. Cincinnati's drinking water Contamination Warning System goes through full-scale exercise. J. Am. Water Works Assoc. 101 (2), 52–55.

Gray, J., 2008. Water contamination events in UK drinking-water supply systems. J. Water Health 6, 21–26.

Guidorzi, M., Franchini, M., Alvisi, S., 2009. A multi-objective approach for detecting and responding to accidental and intentional contamination events in water distribution systems. Urban Water J 6 (2), 115–135.

Haxton, T.M., Uber, J., 2010. Flushing under source uncertainties. American Society of Civil Engineers (ASCE), 12th Annual Water Distribution Systems Analysis (WDSA) Conference, Tucson, AZ.

Hrudey, S.E., Rizak, S., 2004. Discussion of "Rapid analytical techniques for drinking water security investigations." J. Am. Water Works Assoc. 96 (9), 110–113.

Jalba, D., Cromar, N., Pollard, S., Charrois, J., Bradshaw, R., Hrudey, S., 2010. Safe drinking water: Critical components of effective inter-agency relationships. Environ. Int. 36, 51–59.

Kang, D., Lansey, K., 2010. Real-time optimal valve operation and booster disinfection for water quality in water distribution systems. J. Water Res. Plann. Manag. ASCE 136 (4), 463–473.

Lansey, K., Pasha, F., Pool, S., Elshorbagy, W., Uber, J., 2007. Locating satellite booster disinfectant stations. J. Water Res. Plann. Manag. ASCE 133 (4), 372–376.

Lindquist, H.D.A., Harris, S., Lucas, S., Hartzel, M., Riner, D., Rochele, P., DeLeon, R., 2007. Using ultrafiltration to concentrate and detect *Bacillus anthracis, Bacillus atrophaeus* subspecies globigii, and *Cryptosporidium parvum* in 100-liter water samples. J. Microbiol. Methods. 70 (3), 484–492.

Magnuson, M.L., Allgeier, S.C., Koch, B., De Leon, R., Hunsinger, R., 2005. Responding to water contamination threats. Environ. Sci. Technol. 39 (7), 153A.

Moses, J., Bramwell, M., 2002. Champlin Water Works seeks right level of security against terrorist threat. J. Am. Water Works Assoc. 94 (4), 54–56.

Nuzzo, J.B., 2006. The biological threat to US water supplies: Toward a national water security policy. Biosecurity Bioterrorism-Biodefense Strategy Pract. Sci. 4 (2), 147–159.

Parks, S.L.I., VanBriesen, J.M., 2009. Booster disinfection for response to contamination in a drinking water distribution system. J. Water Res. Plann. Manag. ASCE 135 (6), 502–511.

Poulin, A., Mailhot, A., Grondin, P., Delorme, L., Periche, N., Villeneuve, J.P., 2008. Heuristic approach for operational response to drinking water contamination. J. Water Res. Plann. Manag. ASCE 134 (5), 457–465.

Preis, A., Ostfeld, A., 2008. Multiobjective contaminant response modeling for water distribution systems security. J. Hydroinformatics 10 (4), 267–274.

Rice, E.W., Rose, L.J., Johnson, C.H., Boczek, L.A., Arduino, M.J., Reasoner, D.J., 2004. Boiling and *Bacillus* spores. Emerg. Infect. Dis. 10 (10), 1887–1888.

Roberson, J.A., Morley, K.M., 2005. Contamination warning systems for water: An approach for providing actionable information to decision-makers. American Water Works Association, Denver, CO.

States, S., Newberry, J., Wichterman, J., Kuchta, J., Scheuring, M., Casson, L., 2004. Rapid analytical techniques for drinking water security investigations. J. Am. Water. Works. Assoc. 96 (1), 52–64.

States, S., Wichterman, J., Cyprych, G., Kuchta, J., Casson, L., 2006. A field sample concentration method for rapid response to security incidents. J. Am. Water. Works. Assoc. 98 (4), 115–121.

U.S. Army Center for Health Promotion and Preventive Medicine, 2003. Drinking water consumer complaints: Indicators from distribution system sentinels, USACHPPM TG 284. Water Supply Management Program, Aberdeen Proving Ground, MD.

U.S. Environmental Protection Agency (U.S. EPA), 2002. Guidance for water utility response, recovery, and remediation actions for man-made and/or technological emergencies, EPA 810-R-02-001. EPA, Washington, DC.

U.S. Environmental Protection Agency (U.S. EPA), 2003a. Module 1: Water utility planning guide, EPA-817-D-03–001. Response Protocol Toolbox (RPTB) interim final: Planning for and responding to contamination threats to drinking water systems. EPA, Washington, DC.

U.S. Environmental Protection Agency (U.S. EPA), 2003b. Module 2: Contamination threat management guide, EPA-817-D-03–002. Response Protocol Toolbox (RPTB) interim final: Planning for and responding to contamination threats to drinking water systems. EPA, Washington, DC.

U.S. Environmental Protection Agency (U.S. EPA), 2003c. Module 3: Site characterization and sampling guide, EPA-817-D-03–003. Response Protocol Toolbox (RPTB) interim final: Planning for and responding to contamination threats to drinking water systems. EPA, Washington, DC.

U.S. Environmental Protection Agency (U.S. EPA), 2003d. Module 4: Analytical guide, EPA-817-D-03–004. Response Protocol Toolbox (RPTB) interim final: Planning for and responding to contamination threats to drinking water systems. EPA, Washington, DC.

U.S. Environmental Protection Agency (U.S. EPA), 2003e. Module 5: Public health response guide, EPA-817-D-03–005. Response Protocol Toolbox (RPTB) interim final: Planning for and responding to contamination threats to drinking water systems. EPA, Washington, DC.

U.S. Environmental Protection Agency (U.S. EPA), 2004. Response Protocol Toolbox: Planning for and responding to drinking water contamination threats and incidents, response guidelines. EPA, Washington, DC.

U.S. Environmental Protection Agency (U.S. EPA), 2006. Water security handbook: Planning for and responding to drinking water contamination threats and incidents, EPA 817-B-06-001. EPA, Washington, DC.

U.S. Environmental Protection Agency (U.S. EPA), 2007a. Effective risk and crisis communication during water security emergencies: Report of EPA sponsored message mapping workshops, EPA-600-R-0-027. EPA, Washington, DC.

U.S. Environmental Protection Agency (U.S. EPA), 2007b. Water security initiative: Interim guidance on planning for contamination warning system deployment, EPA 817-R-07-002. EPA, Washington, DC.

U.S. Environmental Protection Agency (U.S. EPA), 2008. Decontamination and recovery planning—Water and wastewater utility case study, EPA 817-F-08-004. EPA, Washington, DC.

U.S. Environmental Protection Agency (U.S. EPA), 2011. Planning for an emergency drinking water supply, EPA 600-R-11-054. EPA, Washington, DC.

U.S. Environmental Protection Agency (U.S. EPA), 2012a. Coordination of the water sector and emergency services sectors: An important step to better response, EPA 817-K-12-001. EPA, Washington, DC.

U.S. Environmental Protection Agency (U.S. EPA), 2012b. Laboratory resources for the water sector to support decontamination activities, EPA 817-F-12-003. EPA, Washington, DC.

van der Schalie, W.H., Shedd, T.R., Widder, M.W., Brennan, L.M., 2004. Response characteristics of an aquatic biomonitor used for rapid toxicity detection. J. Appl. Toxicol. 24 (5), 387–394.

Watson, J.T., Gayer, M., Connolly, M.A., 2007. Epidemics after natural disasters. Emerg. Infect. Dis. 12 (1).

Whelton, A.J., Dietrich, A.M., Burlingame, G.A., Cooney, M.F., 2004. Detecting contaminated drinking water: Harnessing consumer complaints. American Water Works Association (AWWA) Water Quality Technology Conference (WQTC), San Antonio, TX.

World Health Organization (WHO), 2002. Environmental health in emergencies and disasters: A practical guide. WHO, Geneva, Switzerland.

World Health Organization (WHO), 2004. Public health response to biological and chemical weapons. WHO guidance. WHO, Geneva, Switzerland.

Rehabilitation

1. INTRODUCTION

Following the confirmation and immediate response to a contamination event, a remediation effort must follow to restore full functionality to the affected system. In extreme contamination cases, it may be necessary to completely abandon the water system and construct a new one. If rehabilitation methods generate waste streams, such as large volumes of wastewater containing the contaminant and other pollutants (e.g., chlorine), the resulting waste streams must be treated prior to disposal. In addition, contamination of drinking water systems can result in the secondary contamination of storm sewers and wastewater systems; these systems may require rehabilitation as well. In this chapter, potential rehabilitation methods are reviewed.

2. REHABILITATION APPROACH AND REGULATORY GUIDELINES

Procedures outlined by the U.S. EPA for rehabilitating water systems and waste streams affected by contamination events are based on Superfund protocols used for hazardous waste site rehabilitation (U.S. EPA, 2003a). The National Oil and Hazardous Substances Pollution Contingency Plan (also known as the *National Contingency Plan*) describes the Superfund remedial response program under the Comprehensive Environmental Response, Compensation, and Liability Act (CERCLA). Using the Superfund model for planning and carrying out a rehabilitation effort is advantageous, as many involved in the rehabilitation process are already familiar with the existing protocols (U.S. EPA, 2003a).

Following a contamination event, external agencies may lead the remediation effort using an incident command system (ICS) (U.S. EPA, 2003a). The ICS is an organizational structure, as defined in a water utility's emergency response plan (ERP), used to facilitate coordination and communication among the water utility, federal, state, and local agencies (U.S. EPA, 2003a). The ICS is led by the incident commander (IC), who is tasked with coordinating efforts from the U.S. EPA, FBI, FEMA, water utility, and other local, state, and federal agencies. The ICS should also designate a water utility emergency response manager (WUERM), who has the authority to coordinate the initial emergency response during a possible contamination event. The WUERM, along with the water utility emergency operations center manager, may be called on as technical advisors for lead external agencies in the ICS. While rehabilitation efforts may be coordinated by external agencies, the involvement of the local water utility staff is vital for success, as they have the most knowledge and experience working with the water system (U.S. EPA, 2003a).

State and local authorities are usually responsible for overseeing contamination remediation; however, depending on the scale and location of the contamination, federal assistance and oversight may be provided if local and state resources are inadequate (U.S. EPA, 2003a). Under the Federal Response Plan (FRP), if national interests are involved, a single federal agency may take the lead and coordinate all other agencies in the remediation effort (U.S. EPA, 2003a). As remediation efforts progress and the scope of the project changes, there will be a shift in the agencies involved, with the U.S. EPA overseeing long-term restoration and cleanup efforts (U.S. EPA, 2003a).

Rehabilitating contaminated water systems involve lengthy, extensive, and expensive processes, including the following steps (U.S. EPA, 2003b):

- Conduct a risk assessment.
- Develop a conceptual site model.
- Conduct system characterization.
  - System characterization work plan.
  - Sampling and analysis plan.
  - Quality assurance project plan.
  - Field sampling plan.
  - Health and safety plan.
- Conduct a feasibility study.
  - Treatability study.
- Conduct a detailed analysis of alternatives.
- Select a remedy.
- Prepare the remedial design.
- Develop a disposal strategy.
- Undertake remedial action.
- Conduct a postremediation monitoring and operations assessment.
- Ensure a long-term alternate water supply.
- Provide public communication.
- Return to normal operation.

3. REHABILITATION PLANNING AND ASSESSMENT

3.1. Risk Assessment

After a contamination event is confirmed, a rapid risk assessment must be performed to determine the possible threat posed to the public and utility workers and help guide remediation efforts (U.S. EPA, 2003a). A risk assessment is also helpful for determining the extent of rehabilitation needed (Haas, 2002). The U.S. EPA or the Federal Emergency Management Agency (FEMA) typically leads the initial risk assessments in conjunction with other agencies, such as the Centers for Disease Control (CDC) (U.S. EPA, 2003a). As potential remediation actions are considered, additional risk assessments may be required to determine the long-term risk reduction of the proposed actions. A simplified bioterrorism risk assessment was developed by Radosavljevic and Belojevic (2009) for quick application and preparation among public services and security professionals.

3.2. Goals

The primary goals for water system rehabilitation are to restore the water supply for fire protection, sanitation, and ultimately for consumption, without compromising the health of the public (U.S. EPA, 2003a). Intermediate goals may help large systems when dealing with concentrated contamination levels. Examples of intermediate goals are treating the water to the level where it can be used for sanitation, firefighting, and other nonpotable uses (U.S. EPA, 2003a). Preliminary remediation goals (PRGs) can be established to facilitate the screening and selection of remediation actions. PRGs can be based on action levels, such as the maximum contaminant level (MCL) for drinking water, or they can be risk-based if no MCL exists (U.S. EPA, 2003a).

3.3. Conceptual Site Model

The development of a conceptual site model (CSM) is a vital part of the planning process, as it facilitates the consolidation of data and helps identify areas where more information is needed (U.S. EPA, 2003a). A CSM is the basis used for further system characterization and should provide information on the properties of the contaminant (from US EPA's Water Contamination Information Tool, WCIT), point of contamination, fate and transport, pathways of exposure, and health risks. CSMs can complement the development of sampling plans as well as risk assessments. GIS maps and hydraulic modeling of the water system are important for determining both the extent of the contamination and the strategy for rehabilitation. Software such as PipelineNet and RiverSpill can assist in GIS and hydraulic modeling (U.S. EPA, 2003a).

3.4. System Characterization

System characterization is a broad and detailed analysis of the nature, extent, and fate of a contaminant in the water system and is typically more in-depth than the initial site characterization (refer to Chapter 6) that takes place during the response phase (U.S. EPA, 2003a). System characterization is meant to fill in information gaps from initial site characterizations and must be customized for each circumstance with the specific purpose of planning remediation efforts; it includes other supporting documents, such as the system characterization work plan, sampling and analysis plan, and health and safety plan (U.S. EPA, 2003a). In system characterization, is it important to identify the limits of the contaminated area to help determine the extent of remediation needed and estimate the required costs and time frame (U.S. EPA, 2003a). Results from the system characterization facilitate the feasibility study, remediation selections, and further risk assessments.

The U.S. EPA recommends the "triad approach" for rapid system characterization. The triad approach is a technical framework for effective cost reduction and remediation streamlining of small contamination events by integrating systematic project planning, dynamic work strategies, and rapid contaminant analysis (U.S. EPA, 2003a). Systematic planning involves a clear, deliberate, and coordinated plan to account for factors (e.g., budget, staff schedules, equipment availability, regulations) that lead to decision uncertainty (U.S. EPA, 2003c). Thorough systematic planning is necessary to reduce costs, facilitate remediation, and reach defensible decisions. The triad approach also utilizes a dynamic work plan that incorporates contingency plans developed in advance, allowing for quick decision making as new data or situations arise (U.S. EPA, 2003c). Data are continuously collected using field-based technologies or rapid laboratory analyses to support correct decision making and rapid responses. The U.S. EPA's Water Laboratory Alliance (WLA) provides a nationwide list of laboratories capable of analyzing water samples, for rapid analysis or confirmation purposes (U.S. EPA, 2012b). The triad approach allows for the blending of system characterization into remediation actions, for a quick, seamless, and defensible response.

3.4.1. Sampling and Analysis Plan

A sampling and analysis plan (SAP) is used to collect data for use with system characterization. The SAP contains two parts, a quality assurance project plan (QAPP) and a field sampling plan (FSP) (U.S. EPA, 2003a).

A QAPP is necessary to ensure the validity and accuracy of reported data and analyses and to ensure the project results meet all specifications. The QAPP is a written document that consist of details about project management, data generation and acquisition, assessment and oversight, and data validation and usability (U.S. EPA, 2003a). Data generation and acquisition, one of the most important parts of the QAPP, should include details about sampling methods and handling, relevant analytical methods, quality control procedures, instrument inspection and maintenance, calibration frequency, and data management (U.S. EPA, 2003a). Quality control samples, such as laboratory duplicates, spiked samples, reagent blanks, and calibration check samples are important for verifying the precision of measurements. Standard operating procedures (SOPs) should be maintained for all relevant analytical methods, and periodic instrument maintenance and calibration should be performed. The QAPP should also contain details on data verification, validation, and the associated methods used. The QAPP should ultimately identify the objectives of the data acquisition, demonstrate that the chosen methods are appropriate, confirm the data gathered is adequate, and establish limitations on the use of the data (U.S. EPA, 2003a). Additional information for QAPP development can be found in U.S. EPA (2002c).

An FSP provides details on sampling methods to be used in conjunction with system characterization. An FSP is more in-depth than initial site characterization plans and should provide details on sample objectives, location, frequency, identification, procedures and methods, handling, analysis, and a system description (U.S. EPA, 2003a). The FSP should also specify the equipment to be used, decontamination procedures, holding times, and sample preservation methods (U.S. EPA, 2003a). A system map showing the locations of sample points should be developed. Quality control samples, such as field duplicates and trip blanks, are necessary to ensure accurate and reproducible sampling methods and should be taken as described in the QAPP. In addition to contaminated water, an FSP must also take into account the possibility of contamination in soils, sediments, deposits in pipes, and the pipes themselves (U.S. EPA, 2003a). Although the FSP should follow the U.S. EPA and other standardized methods for sampling and testing, standardized methods may not be available for the collection of samples from specific drinking water infrastructure surfaces. Methods such as brushing and scraping internal pipe surfaces have been studied (Packard and Kupferle, 2010); however, there is a need for the development of additional sampling methods.

3.4.2. Health and Safety Plan

A health and safety plan (HASP) is required as part of the system characterization, and contains information on site task risk assessment, communication and authority structures, site security, and emergency and medical procedures (U.S. EPA, 2003a). The Occupational Safety and Health Administration (OSHA) is responsible for establishing and implementing safety standards for emergency workers, and under the Hazardous Waste Operations and Emergency Response standard, requires adequate training of workers, providing workers with personal protective equipment (PPE), access to first aid kits, the use of an ICS, and a written HASP (U.S. EPA, 2003a). HASPs should be adjusted for each specific scenario and require proper staff training in using PPE, avoiding common routes of exposure, decontamination procedures, contingency plans, and basic first aid (U.S. EPA, 2003a). An HASP should also address air quality monitoring, personnel monitoring, monitoring techniques and instrumentation, safe handling practices, and safe areas for washing and eating (U.S. EPA, 2003a). Guidance for an HASP can be found in NIOSH (1985), as many of the health considerations for hazardous waste sites are similar for water utilities during a contamination event (U.S. EPA, 2003a).

3.4.3. Feasibility Study

A feasibility study helps in the development, screening, and evaluation of remediation actions. Available remediation actions should be screened based on a set of evaluation criteria, including (U.S. EPA, 2003a):

- Overall protection of human health and the environment
- Compliance with applicable regulations (e.g,. health and safety, environmental)
- Long-term effectiveness and permanence
- Reduction of toxicity or infectivity and mobility through treatment
- Generation of residuals
- Short-term effectiveness
- Implementability and flexibility
- Cost
- State (support agency) acceptance
- Community acceptance

The U.S. EPA's WCIT contains information regarding the appropriate remediation technologies for select contaminants. Additional data on contaminants and their appropriate remediation technologies can be found in the U.S. EPA's Treatability Database. If no existing data are available for the target contaminant, treatability studies need to be conducted to demonstrate the effectiveness of the selected treatment technologies. Treatability studies may be conducted using bench-scale, pilot-scale, and ultimately full-scale treatment systems.

Studies are available that looked at various factors for rehabilitation methods; however, site-specific investigation is still needed. Volchek et al. (2006) reviewed methods for rehabilitating contaminated structures and facilities, some of which are relevant to the rehabilitation of water systems and supporting facilities. Many of the rehabilitation methods and approaches are costly. Selvakumar et al. (2002) provided cost estimates for water distribution pipeline rehabilitation for use with initial planning and budgeting. Of the pipeline replacement techniques reviewed by Selvakumar et al. (2002), microtunneling and horizontal direction drilling were the most expensive options.

4. REHABILITATION METHODS

Rehabilitation of an affected water system depends on the portion of the water system attacked and the nature of the contaminant. Drastically different approaches may be necessary, depending on which of the following portions of the water system is affected: source water, intake structures, water treatment facilities, finished water reservoirs, distribution systems (pipelines, pump stations, etc.), or household plumbing. Soils, storm sewers, or sanitary sewers may also be affected, depending on the source of the contamination. The general rehabilitation approach depends on whether a biological, chemical, or radiological contaminant is present. Long-term rehabilitation strategies following a contamination event require cleaning, repairing, or replacing the affected system and determining whether to use the existing water source or secure a new source

Cleaning distribution systems is a regular part of water system operations. Over time, water system components, such as pipes, reservoirs, and pump stations, develop biological growth (referred to as *biofilms*), chemical scaling, and deposits of particulate matter. In addition, corroded pipes can adsorb and retain biological contaminants that persist (Szabo et al., 2007). During a contamination event, it is possible that the contaminant can become associated with biofilms and solids in the water system. Cleaning methods for distribution systems that are used to remove contaminated water, such as flushing and pigging, can also remove the sediments and biofilms that might contain the contaminant.

4.1. Treatability of Contaminants

The most applicable treatment strategies depend on the properties of the contaminant. For evaluating the most beneficial treatment, it is best to determine the following characteristics of the contaminant (U.S. EPA, 2003b):

- Hydrolysis
- Reactivity, including biodegradability
- Solubility

- Implications of oxidation
- Volatilization
- Stability in water
- Susceptibility to disinfection

In addition to the chemical and biological characteristics of contaminants, it is beneficial to determine the following health effects information, which influences how the contaminant is treated and the level of treatment required:

- Acute health effects
- Chronic health effects
- Toxicity values, such as median lethal dose (LD_{50})
- Relationship between morbidity and mortality
- Secondary transmission characteristics

4.2. Cleaning Methods

The American Water Works Association (AWWA) established general guidelines for cleaning methods used for distribution systems such as flushing, cleaning, disinfecting, and rehabilitating water system components. Relevant guidance manuals include:

- AWWA Manual M17: *Installation, Field Testing, and Maintenance of Fire Hydrants (AWWA, 2006)*
- AWWA Manual M14: *Recommended Practice for Backflow Prevention and Cross-Connection (AWWA, 2004)*
- AWWA Standard C564-87: *Disinfection of Wells (AWWA, 2013b)*
- AWWA Standard C651-05: *Disinfecting Water Mains (AWWA, 2005)*
- AWWA Standard C652-11: *Disinfection of Water-Storage Facilities (AWWA, 2011)*
- AWWA Standard C653-03: *Disinfection of Water Treatment Plants (AWWA, 2013a)*

Additional guidance is available on cleaning distribution system components (Ainsworth, 2004; Kirmeyer, 2000; Kirmeyer et al., 2001; U.S. EPA, 2002a). Table 7.1 contains a list of alternatives that could be used to clean water system components, most of which are also used to clean distribution systems under normal operations.

4.3. Flushing

Flushing is the most common method used to maintain water quality in drinking water systems, and as a result, the water industry is highly experienced with flushing. Successful flushing can result in removal of accumulated silt and sediment, reduction of chlorine demand, removal of accumulated biofilms, and removal of contaminated water (Friedman et al., 2002). The three types of flushing are

TABLE 7.1 Cleaning Methods for Water System Infrastructure Following a Contamination Event

Rehabilitation Method	System Components	Description
Flushing	Pipelines	Clean water is continuously introduced and withdrawn to capture and remove contaminants, specifically those associated with particulates and adhered to pipe walls. Sufficient flushing velocities must be developed (e.g., 2.0–2.5 ft/sec). Flushing requires large volumes of clean water and disposal of contaminated effluent water.
Power washing	Tanks Reservoirs Pump stations Canals	Clean water is used to clean surfaces and remove deposits by the use of water pressure. This requires large volumes of clean water and disposal of contaminated effluent water. It may require personnel to enter enclosed spaces where contaminants are present.
Chemical disinfection	Pipelines Pump stations Wells Treatment plants Canals Reservoirs	A solution containing a high dose of chemical disinfectant (typically chlorine) is added to the affected portion of the water system for a time period to achieve the desired inactivation of the target microorganisms. Flushing must follow chemical disinfection to ensure removal of the disinfectant to acceptable levels. This requires large volumes of clean water and disposal of contaminated effluent water that also contains the disinfectant.
Chemical treatment	Pipelines Pump stations Wells Treatment plant tanks Canals Reservoirs	Solutions containing acids, caustic chemicals, or other chemicals are added to the affected portion of the water system. Flushing must follow chemical treatment to ensure removal of the chemicals to acceptable levels. This requires large volumes of clean water and disposal of contaminated effluent water that also contains the chemicals.
Air scouring	Pipelines	Injection of compressed air into the system is useful when flushing cannot be performed due to low system pressure. Compressed air can also be used during flushing. Air scouring is used to remove adherent materials such as biofilms and other pipe deposits. Materials removed by air scouring must be flushed from the system.
Pigging and swabbing	Pipelines	Pigging involves inserting a bullet shaped object in the pipe and driving it through the pipe using water pressure. Swabbing is similar but the object inserted is of a different shape and consists of a foam sponge, whereas a "pig" can be made of various materials.
Shot blasting	Pipelines Treatment plant tanks Reservoirs	Utilizes very small bullet-shaped objects made of various materials (e.g., polyurethane) to remove pipe deposits and contaminants in the water distribution system, while improving hydraulic flow.

traditional, unidirectional, and continuous blowoff. Traditional flushing consists of using fire hydrants to inject and extract water to achieve sufficient velocities to suspend and remove pipe deposits. Unidirectional flushing involves the use of fire hydrants to remove water as well as isolation valves to isolate the area being flushed (Poulin et al., 2010). Continuous blowoff flushing involves forcing a flow through a section of pipe. Continuous blowoff is used for utilities with numerous dead ends or circulation problems, and it can be helpful for maintaining disinfectant residuals (Friedman et al., 2002). Periodic spot flushing, similar to continuous blowoff flushing, is another method used to address systems with a lot of dead ends and has been proven to reduce the described bacteria growth (Selvakumar et al., 2002). During flushing, pipe velocities must be sufficient to provide a scouring of adhered solids and resuspension of settled solids. Flushing is relatively inexpensive compared to other methods, but the use of flushing produces large quantities of waste wash water, especially if injection locations are not properly identified. Therefore, consideration must be taken when using flushing as well as with any other rehabilitation technique.

Where flushing is used, water system modeling tools should be utilized to select the hydrants to use and assure proper flushing velocities are achieved (Alfonso et al., 2010; Haxton and Uber, 2010; Poulin et al., 2008). Modeling is needed to identify the source of the contamination, the extent of the contaminated area, and an appropriate flushing strategy (e.g., which fire hydrants should be opened). If proper injection locations are not located, flushing may cause the contaminant to spread to other parts of the distribution system. Knowledge of the contaminant injection location greatly influences the effectiveness of flushing (Haxton and Uber, 2010). A multi-objective optimization technique was developed by Alfonso et al. (2010) to provide a fast and effective flushing response to contamination events. The water distribution system modeling software packages WaterCAD and WaterGEMS produced by Bentley (Exton, PA) include modules for simulating flushing in distribution systems.

4.4. Power Washing

Power washing is a cleaning method used in reservoirs, tanks, and pump stations to clean internal surfaces and rid them of accumulated materials adhering to the surface and deposits (Godfrey, 2005). Infrastructure components must be taken out of service and drained prior to power washing. Power washing requires the use of specialized equipment that ejects water at high pressures. Proper equipment must be selected to reduce the risk of damaging weak surfaces and coatings, potentially exposing aggregate on concrete surfaces due to the high pressure or damaging reservoir liners (Ainsworth, 2004). Power washing may be followed by disinfection. A large volume of waste wash water is generated as a result of power washing, and this waste stream must be treated prior to disposal. Another concern with using power washing is that workers must enter the reservoir to clean it, which may expose them to contaminants within the reservoir.

In lieu of taking a reservoir out of service for power washing, water providers can elect to clean reservoir surfaces while the system is still in operation. One way to accomplish this is to use specially trained commercial divers and cleaning equipment that operates using a vacuum. Alternatively, automated or remotely controlled equipment can be used in place of divers. Water quality in the reservoir is usually not compromised during cleaning using this method. Similar to power washing, the generated waste wash stream must be treated prior to disposal. Unmanned equipment would be advantageous in a contamination event because it would limit the exposure of personnel to the contaminants. As part of this study, two companies were located that manufacturer equipment for unmanned cleaning of tanks and reservoirs: VoR Water VR systems and Acuren.

4.5. Disinfection

Disinfection is another method used to clean water distribution systems and ensure water quality. Chemical disinfectants used include free chlorine, combined chlorine, and chlorine dioxide (U.S. EPA, 2003b). Chemical disinfection consists of preparing a strong disinfectant solution and introducing it into the water system for a prescribed period of time. The efficacy of inactivating microorganisms is a function of the disinfectant residual concentration and the contact time. Following the disinfectant contact time, the distribution system components must be drained and then flushed. In addition, flushing or power washing may be advisable prior to disinfection to reduce the chlorine demand of system components. Because powders and granules are often used as disinfectants, adequate mixing is important and appropriate measures should be taken to introduce chemical disinfectants into tanks and pipelines. The WHO provides guidelines for disinfecting tanks and tank trucks, including recommended disinfectant concentrations and contact times (Godfrey, 2005).

Chlorine disinfection is effective for only certain (e.g., biological) contaminants. Some microorganisms (e.g., the oocysts of *Cryptosporidium*) are resistant to inactivation by chlorine. Hosni et al. (2009) studied the inactivation of *Bacillus globigii* spores, a surrogate for anthrax, using both chlorine dioxide and chlorine, and found chlorine dioxide required much shorter contact times for deactivation than chlorine. Despite this fact, the concentrations of chlorine dioxide required for adequate *Bacillus globigii* inactivation would render the water nonpotable, and disinfection would have to occur in batch mode with flushing (Hosni et al., 2011). Alternatively, Szabo et al. (2012) investigated the intentional germination of *Bacillus globigii* spores followed by inactivation with free chlorine and found a dose of 5 mg/L reduced spore levels to undetectable levels after four hours. Kauppinen et al. (2012) studied the inactivation of *E. coli* and adenovirus using peracetic acid and chlorine and found that, while both could effectively treat *E. coli* and adenovirus, adenovirus was not inactivated as efficiently, implying virus levels may be underestimated if *E. coli* is used as a water quality parameter. Overall efficiency of disinfection chemicals can vary greatly depending on the target microbe, concentration of spores or cells, decontaminant characteristics, decontaminant concentration, and exposure time (Raber and Burklund, 2010). Note that, for corroded pipes, biological agents may be protected from disinfectants; Szabo et al. (2007) found that *Bacillus globigii* spores injected into a simulated distribution system were present in corroded pipes following disinfection.

An additional disinfection method used in water treatment is ultraviolet (UV) disinfection, although no sources documenting its use in water distribution systems to inactivate microorganisms on surfaces were located. Regardless, UV systems have the potential to be an effective point of use disinfection system for water with no disinfectant residual. UV light has been studied for inactivation of microorganisms on hospital surfaces and in patient rooms. In one study it was found that automated, portable UV light significantly reduced aerobic colony counts on high-touch surfaces in patient rooms, in particular on surfaces that were exposed directly to UV light (Boyce et al., 2011). Rose and O'Connell (2009) studied the inactivation of seven possible bacterial contaminants using UV light and found gram-negative bacteria required less than 12 mJ/cm^2 to achieve a 4-log$_{10}$ inactivation, while *B. anthracis* spores required 40 mJ/cm^2 for only a 2-log$_{10}$ inactivation.

4.5.1. Effect of Biofilms and Scaling

Rehabilitation of contaminated systems is made difficult by the presence of biofilms and scaling on the surfaces of pipes and other water infrastructure components (Porco, 2010). Introduced microbial contaminants may become integrated into biofilms attached to the system component surfaces. Biofilms can and should be controlled in distribution systems as part of normal operations, due to microbial contaminants present in biofilms being particularly difficult to inactivate (Mains, 2008). Vitanage et al. (2004) provided guidance on the maintenance and design of distribution systems to minimize biofilm growth and make cleaning and rehabilitation easier tasks. Successful management of biofilms includes regular flushing, line pigging, line replacement, and maintaining adequate levels of residual chlorine (Mains, 2008).

Experimental work indicates that the disinfection of microorganisms occluded in drinking water system biofilms is inhibited by both biofilm structure and microbial resistance to the disinfectant (Morrow et al., 2008; Szabo et al., 2007). Studies using *Bacillus* spores associated with biofilms found the required disinfectant concentrations to be 5–10 times higher than for free floating spores (Morrow et al., 2008). Biofilms of *Pseudomonas aeruginosa* and *Klebsiella pneumoniae* were found to be resistant to both penetration and deactivation by hypochlorite, and while chlorine-based biocides could penetrate the biofilms, deactivation remained poor (Stewart et al., 2001). In addition, pipe materials that have rough or corroded surfaces support fast growing biofilms, and the microbes within are relatively inaccessible to disinfectants, meaning they must be removed using physical processes, sloughing, or shearing (Szabo et al., 2006,

2007). In addition to microorganisms, biofilms have also been shown to inhibit the effects of flushing, free chlorine, and ammonia on the decontamination of cesium chloride, a radioisotope surrogate (Szabo et al., 2009).

4.6. Chemical Treatment

Chemical treatment is another method used to address distribution system components and contamination. Similar to chemical disinfection, chemical treatment involves injection of acids and bases into the distribution system followed by flushing. A wide range of chemicals may be used and each chemical should be selected for certain applications. Engineers need to confirm that the chosen chemicals are suitable for a potable water system and do not harm pipes and other structures (Ainsworth, 2004).

4.7. Air Scouring

Air scouring can be used in pipelines where flushing cannot be performed due to low system pressure or large pipe diameters (U.S. EPA, 2003b). Air scouring is performed by isolating a section of the system, injecting filtered compressed air into the line, and collecting the air/water mixture exiting the line (U.S. EPA, 2003b). The water/air mixture, called a *slug*, is driven by the compressed air at high velocity (Ainsworth, 2004; Kitney et al., 2001). The setup for air scouring consist of three portions, the compressor, air cooler, and filtering system (Kitney et al., 2001). Following air scouring, the exiting air/water mixture is collected and disposed of properly. In addition, it is important to remove all the compressed air from the pipe before returning to service to avoid unstable flows and cloudy water (Ainsworth, 2004). One major advantage of air scouring is that it requires 40% less water than flushing and swabbing (Ainsworth, 2004).

4.8. Pigging and Swabbing

Pigging and swabbing are two pipeline cleaning techniques that involve introducing bullet-shaped objects into pipelines through hydrants or insertion points (Ainsworth, 2004; Quarini et al., 2010). As the pigging and swabbing objects travel through the pipe, biofilms, scaling, and other accumulated materials are disrupted and pushed through the pipe (U.S. EPA, 2003b). A "pig" can be made of food-grade silicon, foam, steel, plastic, or polyurethane; and these pigs range in size from 2 to 48 inches in diameter and can have various lengths, styles, and configurations, depending on the pipe material (Satterfield, 2007). A pig is inserted in the pipeline through a launcher, then clean water is introduced to move the pig forward (Satterfield, 2007). Following pigging, pipelines must be flushed to remove materials removed from the pipe surfaces during pigging. Flushing results in a large waste stream that must be treated.

Swabbing is similar to pigging except a different type of object is used. For swabbing the objects are identified as *swabs* instead of *pigs*, and the swab is usually made of foam. The diameter of the swab is approximately 25% greater than the pipe it is being forced through (Ainsworth, 2004). Swabs can be classified as soft, hard, or scouring; and the choice of swab depends on the pipe material (Ainsworth, 2004). Swabs are capable of removing soft deposits, and it is typical to send between three and six swabs through a pipe to achieve adequate cleaning (Ainsworth, 2004). Swabbing is most effective at velocities ranging from 0.8 to 1.5 m/s (Ainsworth, 2004). Similar to the case for pigging, flushing must follow swabbing, which generates a large waste stream.

4.9. Shot Blasting

Shot blasting is a cleaning method used where media is fired at the component surface. Shot blasting is used in pipelines, treatment plant tanks, and reservoirs. The media chosen for shot blasting can vary in aggressiveness and typically consists of sodium bicarbonate, sand, or bullet-shaped polyurethane (U.S. EPA, 2003b). If shot blasting is done correctly, it can remove most accumulated contaminants, sediments, soft scales, biofilms, and other impurities that have been deposited on the surface (U.S. EPA, 2003b). As with other rehabilitation methods that utilize high pressure and force, care must be taken in choosing the correct equipment and media to prevent damage to the component surface.

5. ADDITIONAL REHABILITATION METHODS

In cases where the cleaning of water distribution system components is not sufficient to remove contaminants and return the system to service, it may be necessary to take more significant measures to rehabilitate system components. In the most extreme cases of contamination, removal and replacement of system components may become necessary. Table 7.2 contains a description of additional rehabilitation methods.

5.1. Liner Systems

Nonstructural liner systems are utilized in pipelines, pump stations, storage tanks, and reservoirs. AWWA Standard D102 recognizes general types of interior coating systems, including epoxy, vinyl, enamel, and coal-tar (U.S. EPA, 2002b). Although all the coatings described by AWWA can be used, cement and epoxy resin are the most utilized (Ainsworth, 2004). The application of liners may be accomplished in one of two ways, by spray or by brush (Ainsworth, 2004). Applied liners are thin corrosion-resistant material that prevent leaks and increase service life (Selvakumar et al., 2002). Additional advantages of applied liners are increased water quality, increased protection of pipe material, and a cost of one third to one half of pipe replacement (U.S. EPA, 2003b).

TABLE 7.2 Additional Rehabilitation Methods for Water System Infrastructure

Rehabilitation Method	System Components	Description
Liner systems	Pipelines Pump stations Storage tanks Reservoirs	Replacement of the liner systems is an alternative to replacement of system components, making it more cost effective. Liner systems may be composed of cement, plastic, and epoxy. For pipelines, it is possible to use cast-in-place plastic piping, where flexible plastic liners are pulled through pipelines and cured to form a layer adhering to the pipe wall.
Sliplining	Pipelines	This process involves pulling or pushing a new pipe, of smaller diameter, into the existing pipe and grouting the annular area between the pipes to prevent leakage and provide structural integrity.
Pipe bursting	Pipelines	Involves breaking open the existing pipe and simultaneously pulling or push a new pipe in place behind a bursting head.

5.2. Sliplining

Trenchless technology has been utilized for pipeline rehabilitation for many years. Advantages of trenchless technology include reduced surface disruptions, lower reinstallation costs, and shorter construction times (O'Reilly and Stovin, 1996). Risk assessment and cost savings associated with employing trenchless technology are discussed by O'Reilly and Stovin (1996) and Tighe et al. (2002), respectively. One of the oldest and most inexpensive trenchless technology methods is sliplining (Selvakumar et al., 2002). Sliplining involves pulling or pushing a new pipe of smaller diameter into an existing pipe and grouting the annular area between the pipes to prevent leakage and provide structural integrity. The maximum diameter of the smaller pipe is usually about 10% smaller than the existing pipe to allow for easy installation (Zhao and Daigle, 2001). Sizes of existing pipes suitable for sliplining can range from 4 to 108 inches. Liner pipes (new pipes) are usually made of either high-density polyethylene (HDPE) or fiberglass-reinforced polyester (Selvakumar et al., 2002). When designing for slipline rehabilitation, it is usually common practice to neglect the structural contributions of the existing host pipe and grout (Zhao and Daigle, 2001). However, an engineer needs to check that the capacity of the new pipe is not exceeded due to a reduced cross-sectional area and smoother pipe material (Najafi and Gokhale, 2005). TAG-R Online is a fully automated web-based decision support system for the selection of underground utility construction methods, focusing on trenchless technology (www.tagronline.com).

5.3. Pipe Bursting

Pipe bursting is another form of trenchless technology used in pipe rehabilitation. Pipe bursting involves breaking open the existing pipe while simultaneously pulling or pushing a new pipe in place behind the bursting head (Selvakumar et al., 2002). Using pipe bursting, the replacement pipe can be the same size or up to two pipe sizes larger than the existing pipe and, therefore, does not decrease the capacity of the pipe (Selvakumar et al., 2002). Pipe bursting utilizes static, pneumatic, or hydraulic pipe bursting tools, which are drawn through the pipe by a winched cable (Selvakumar et al., 2002). Pipe bursting can be used for pipe diameters between 6 to 48 inches and the replacement pipe is usually made of polyethylene or polyvinyl chloride (PVC) (Selvakumar et al., 2002). Soil conditions and the stiffness of the pipe must be taken into consideration to determine vertical and horizontal pipe deflections during pipe bursting (Lapos et al., 2007).

6. REHABILITATION OF BUILDINGS

Contamination of buildings should be limited to the building's plumbing system and areas immediately surrounding water fixtures, such as bathrooms, kitchens, and laundry rooms. The U.S. EPA recommends two basic steps for building plumbing restoration (U.S. EPA, 2012c):

- Purge contaminated water from the system.
- Flush or treat to remove accumulated contamination.

The easiest and most cost effective method for removing accumulated contamination is continuous flushing of the system with water directly from the distribution system. If water from the distribution system is the source of the contamination, another water source, such as a tanker truck, must be used. Water used for flushing can be heated or amended with additional chlorine, surfactants, or germinant solutions to promote germination and deactivation of residual biological contaminants (U.S. EPA, 2012c). Multiple flushing methods can be implemented, including high-velocity pumping, steam injection, pulsating flow, flooding and letting it stand, and conventional flushing (U.S. EPA, 2012c). Effluent from flushing processes can either be collected for disposal or discharged into the sewer, depending on the contaminant present. The time frame for effective treatment using flushing can vary on the order of days and depends on the design of the plumbing system, type and initial concentration of contaminant, and acceptable residual levels (U.S. EPA, 2012c).

Following flushing, further remediation of plumbing fixtures and associated areas is recommended. Water tanks, including hot water heaters, may require draining and removal of sediments, as it is difficult to effectively flush them due to their large volumes (U.S. EPA, 2012c). Sinks, tubs, and other bathroom and kitchen surfaces should be thoroughly cleaned and faucets, valves, aerators, drains, and hoses may need to be replaced (U.S. EPA, 2012c). Other appliances that use water, such as dishwashers and laundry machines may be cleaned through operation or disconnected and cleaned offline. Appliances that are indirectly exposed to the contaminated water, such as dryers, may also need to be cleaned. Biological contaminants can typically be effectively disinfected using liquid cleaners containing bleach, hydrogen peroxide, or hypochlorous acid/hypochlorite, as well as fumigation techniques that use chlorine dioxide, hydrogen peroxide, or methyl bromide (U.S. EPA, 2011). Decontamination products such as foaming agents produced exclusively for military use are effective on chemical contaminants on nonporous surfaces but are less effective on porous surfaces and can corrode indoor surfaces in the process (Love et al., 2011). Porous surfaces are typically harder to decontaminate than nonporous surfaces and may require replacement. Contamination with radioactive species or highly toxic chemical compounds may also require the complete replacement of contaminated surfaces.

Outdoor remediation may be necessary, depending on the use of sprinkler systems or garden hoses. However, because most people spend almost 90% of their time indoors, priority should be on indoor remediation, as exposure times are significantly greater (U.S. EPA, 2011). Owning pets or young children may increase the priority of outdoor remediation. Outdoor surfaces with ample sunlight may be candidates for natural attenuation, depending on the properties of the contaminant. However, contaminants such as radioactive species or persistent chemical compounds may require the complete removal of contaminated surfaces and top soil to eliminate any risk of exposure. As with other remediation efforts, fate and transport play an important part in deciding appropriate remediation techniques.

7. TRANSPORT AND CONTAINMENT OF CONTAMINATED WATER

Containment of large volumes of water collected during contamination involves a lot of effort (U.S. EPA, 2012a). If the volume of contaminated water is large, treatment of the contaminated water most likely needs to occur on-site. If the volume of contaminated water is relatively small, it can be transported to a storage site, treatment plant, or another site where it can be more easily treated.

An internet search of companies that rent liquid storage equipment revealed the following:

- Tank trucks available for rent typically have capacity for 5000 to 9000 gallons.
- Readily available polyethylene tanks have capacity up to approximately 7000 gallons.
- It is possible to construct lagoons with multiple (e.g., double) liners using readily available supplies.

This information provides some insight into how contaminated water would be handled during a contamination event; however, the numbers are daunting. Even mid-size water systems may have the capacity for millions of gallons of storage within their systems, which may become contaminated during an event. As an example, 1000 feet of 8 inch diameter pipe contains approximately 2600 gallons of water.

8. TREATMENT OF CONTAMINATED WATER

8.1. Water Treatment Technologies

In the event of contamination, the contaminated water needs to be removed from the water system and treated prior to disposal. In addition, it is necessary to treat waste streams generated as the result of rehabilitation and recovery efforts, such as flushing, pigging, or disinfecting of pipelines or cleaning of reservoirs. It is unlikely that a water provider would opt to use the contaminated water as a drinking water source, even if it is treated, although it may be necessary if no other sources are available. Numerous treatment technologies are available, and the selection of the most appropriate treatment depends on the contaminant involved. The treatment technologies shown in Table 7.3 can be applied individually or in combination to increase overall effectiveness. The password-protected U.S. EPA's WCIT contains information on appropriate treatment technologies for many of the contaminants of concern (U.S. EPA, 2007). The database has restricted access, and security clearance is required prior to gaining permission to use it. Additional guidance on drinking water contaminants, properties, fate and transport, and possible treatment technologies can be found in the U.S. EPA Drinking Water Treatability Database (TDB). The TDB contains over 25 treatment technologies and more than 50 chemical, radiological, and biological contaminants, with plans to expand the database to over 200 contaminants. A summary of chlorine disinfection information is also available for a variety of the biological contaminants of concern (Burrows and Renner, 1999).

The treatment processes described in Table 7.3 represent a wide range of technologies that can be used for removal or transformation of many contaminants. For example, chemical treatment is a broad category. Chemical oxidants include oxygen, chlorine, chlorine dioxide, hydrogen peroxide, ozone, and permanganate. Chemical reactions can also be promoted using photolysis.

TABLE 7.3 Water Treatment Technologies (Crittenden et al., 2012)

Process	Contaminants Targeted	Removal Mechanism
Chemical treatment	Inorganic and organic chemicals	Chemical oxidation and reduction processes are used to convert chemical species into less harmful products or precipitate contaminants for easier removal.
Coagulation or flocculation	Inorganic and organic particles, colloidal and dissolved organic matter	Coagulation alters the chemical composition of particulate matter, changes the size distribution of the particles, and captures some of the dissolved and colloidal matter into larger particles so that it can be removed in sedimentation or filtration systems.
Gravity separation	Inorganic and organic particles	In sedimentation basins, particles settle by gravity in a quiescent basin. Dissolved air flotation systems are used to separate solids by floating solids on the water surface.
Granular filtration	Inorganic and organic particles	Particles are retained on the surface of the media and captured in the void space.
Membrane filtration	Inorganic and organic particles	Particles are excluded based on size as water passes through a synthetic material.
Disinfection	Microorganisms	Microorganisms are removed or inactivated using a variety of disinfectants, such as free and combined chlorine, chlorine dioxide, UV light, and ozone.
Air stripping	Volatile organic chemicals	Chemicals are removed by mass transfer from water to air. Air containing the chemicals must be removed.
Adsorption	Organic and inorganic chemicals	Chemicals are removed by mass transfer from water to solid. Solids containing the chemicals must be removed.
Reverse osmosis	Dissolved chemicals	Chemicals are removed by pressure-driven membrane and preferential separation.

Advanced oxidation processes involve production of radicals that aggressively react with chemical species. Advanced oxidation agents include:

- Hydrogen peroxide/UV light
- Hydrogen peroxide/ozone
- Titanium dioxide/UV light
- Ozone/UV light
- Ozone/UV light/hydrogen peroxide
- Ozone
- Fenton's reagent involving iron and hydrogen peroxide or ozone

Emerging technologies for advanced oxidation include those that use sonolysis, supercritical water oxidation, titanium oxide with ozone or ozone and hydrogen peroxide, electron beam radiation, gamma radiation, and electrohydraulic cavitation. Technologies for these emerging processes are not necessarily readily available.

8.2. Mobile Treatment Systems

In some cases, it may be necessary to obtain and use portable commercial treatment systems to treat the contaminated water (Appendix F). These commercial systems can treat water at a rate of up to 40 gpm. Potential power sources vary; some available mobile treatment units operate on solar power. The smaller units can be carried while the larger units can be air lifted or trailer-hitched to the affected area. According to the manufacturer websites, these types of systems are used by the military in response to natural disasters, such as Hurricane Katrina, and overseas in Afghanistan and Iraq. For mobile treatment systems to be used in the event of an emergency, water providers must have ready access to these systems for quick mobilization. Planning for use of mobile treatment systems should be considered for inclusion in the feasibility study as part of the operational response plan (Chapter 6).

The comprehensive mobile treatment systems typically employ a multistage treatment process which begins with filtration and ends with UV disinfection. The filtration process is made up of three or four steps, depending on the manufacturer, and begins with the removal of large debris, sand, grit, and sediment. The final filtration step involves membrane filtration to remove particles with diameters as small as 2 μm. Following filtration, the water is subjected to UV disinfection, which destroys microbial contaminants. These mobile treatment systems are designed primarily to treat poor-quality water from streams or lakes for use as drinking water. These systems are not specifically designed to address removal of contaminants from drinking water in a contamination event. For example, in the case of contaminated drinking water, it is unlikely that there would be a lot of particulate matter that would require removal. In addition, the mobile treatment systems may not adequately address the type of contamination present.

If high-quality drinking water is desired, it is possible to use reverse osmosis (RO), which provides the best available treatment (Burrows and Renner, 1999;

U.S. EPA, 2001). However, treatment of water using RO produces a large volume of brine that requires disposal. If mobile treatment systems are being used to remove contaminants prior to disposal, this type of treatment may not be appropriate.

9. DISPOSAL ISSUES

Contamination events and remediation efforts may lead to the generation of contaminated surface water, ground water, remediation waste streams, and soil and sediments. Personal protective equipment and consumer equipment, such as hoses, toilets, and coffee makers can also become contaminated through public water use and must be cleaned or disposed of accordingly (U.S. EPA, 2003a).

Any contaminated solid or liquid wastes must be managed according to applicable federal, state, or local regulations. Wastes containing contaminants regulated under Resource Conservation and Recovery Act (RCRA), the Clean Water Act (CWA), or the Toxic Substance Control Act must be handled and treated accordingly (U.S. EPA, 2003a). Wastes contaminated with radioactive materials may be subject to regulation by the Nuclear Regulatory Commission, the Department of Energy, and other agencies. Methods for disposal of contaminated water include discharge into surface water or a publicly owned treatment works (POTW) or underground injection.

Disposal of contaminated water and remediation waste streams typically involve treatment and discharge into a POTW or surface water. These processes are regulated under the CWA, and it may not always be possible to discharge even treated water into a POTW if pretreatment standards and appropriate permits are not met (U.S. EPA, 2003a). Another option for disposal may be underground injection. Injection is subject to regulation by the U.S. EPA as part of the Underground Injection Control Program.

The disposal issue is particularly problematic in cases where multiple agencies are involved. The practicality of dealing with decontamination is daunting and overwhelming; dealing with issues such as the disposal of large volumes of cleanup water and expedited permitting issues become important in rehabilitation of a contaminated system (WSDWG, 2008).

10. POSTREMEDIATION MONITORING

Long-term monitoring is needed to confirm the effectiveness of response and remediation efforts before any remediation treatments can be stopped. Postremediation monitoring may require the following actions (U.S. EPA, 2003a):

- Monitoring for contaminants.
- Inspection and maintenance of treatment equipment.
- Inspection and maintenance of water distribution system.
- Maintenance of security measures.
- Public communication of monitoring activities.

After the water system is returned to full operation, long-term monitoring is needed for continual assurance that remediation objectives can be sustained. Sampling and monitoring should occur periodically at various locations throughout the water system. Long-term monitoring can be an important component in ensuring public confidence in the water supply after an incident.

11. LONG-TERM ALTERNATE WATER SUPPLY

During the recovery and rehabilitation process, an interim water supply strategy is needed for drinking water, sanitation, and firefighting (WHO, 2002). During the immediate response, a short-term alternative water supply most likely will be used. During rehabilitation, it may be possible to use boil notices, or point-of-use purification may be deemed adequate to reduce the contamination to levels safe for public health. In these cases, no alternative water supply is needed. If only a restriction on human consumption is in place (the water can be used for sanitation and firefighting), an alternative water supply is needed for drinking water and food preparation purposes. A "do not use" restriction requires an alternate water supply for all water uses (U.S. EPA, 2003a). Multiple options exist for securing alternate water supplies. Mutual aid agreements can be made with surrounding water utilities, contractors, and companies to secure a water source. Supplies including spare pumps, pipes, tanks, and valves may be of value for transporting water to affected areas. Other options include using tank trunks or buying bottled or bulk water from private companies (U.S. EPA, 2003a).

CONCLUSIONS

Following the confirmation and immediate response to a contamination event, a remediation effort must follow to restore full functionality to the affected system, or the system may be abandoned. Rehabilitation is a lengthy, expensive process, involving risk assessments, site models, system characterization, sampling plans, and feasibility studies before action is taken. U.S. EPA procedures for rehabilitating contaminated water systems events are based on Superfund protocols, and cleanup may be led by federal, state, or local authorities.

Rehabilitation methods can vary greatly, depending on the characteristics of the contaminant and the portion of the water system affected. Affected areas can include the source water, water treatment facilities, household plumbing, and storm sewers. Guidelines for selecting appropriate and cost-effective strategies for the rehabilitation of water system components, such as flushing, scouring, or sliplining, have been established by groups such as the American Water Works Association. Rehabilitation methods may also generate contaminated waste streams that must be treated prior to disposal. Methods for treating contaminated wastewater can be found in the U.S. EPA's WCIT and TDB.

During rehabilitation, long-term alternative water supplies may be needed to supply water for drinking, firefighting, or sanitation. After water systems have been rehabilitated, long-term monitoring, inspection, and maintenance are vital to ensure the safety of the water supply and renew public confidence.

REFERENCES

Ainsworth, R., 2004. Safe piped water: managing microbial water quality in piped distribution systems. IWA Publishing, London, UK.

Alfonso, L., Jonoski, A., Solomatine, D., 2010. Multiobjective optimization of operational responses for contaminant flushing in water distribution networks. J. Water Res. Plann. Manage. ASCE 136 (1), 48–58.

American Water Works Association (AWWA), 2004. Recommended practice for backflow prevention and cross-connection, AWWA Manual M14. AWWA, Denver, CO.

American Water Works Association (AWWA), 2005. Disinfecting water mains. AWWA Standard C651-05. AWWA, Denver, CO.

American Water Works Association (AWWA), 2006. Installation, field testing, and maintenance of fire hydrants, Manual M17. AWWA, Denver, CO.

American Water Works Association (AWWA), 2011. Disinfection of water-storage facilities. AWWA Standard C652-11. AWWA, Denver, CO.

American Water Works Association (AWWA), 2013a. Disinfection of water treatment plants. AWWA Standard C653-13. AWWA, Denver, CO.

American Water Works Association (AWWA), 2013b. Disinfection of wells. AWWA Standard C654-13. AWWA, Denver, CO.

Boyce, J.M., Havill, N.L., Moore, B.A., 2011. Terminal decontamination of patient rooms using an automated mobile UV light unit. Infect. Control. Hosp. Epidemiol. 32 (8), 737–742.

Burrows, W.D., Renner, S.E., 1999. Biological warfare agents as threats to potable water. Environ. Health. Perspect. 107 (12), 975.

Crittenden, J.C., Trussell, R., Hand, D.W., Howe, K.J., Tchobanoglous, G., 2012. MWH's Water Treatment: Principles and Design, third ed. John Wiley & Sons, Hoboken, NJ.

Friedman, M., Kirmeyer, G.J., Antoun, E., 2002. Developing and implementing a distribution system flushing program. J. Am. Water. Works. Assoc. 94 (7), 48–56.

Godfrey, S., Reed, B., 2005. Cleaning and disinfecting water storage tanks and tankers. World Health Organization (WHO), Water, Engineering and Development Centre, Leicestershire, UK.

Haas, C.N., 2002. The role of risk analysis in understanding bioterrorism. Risk. Anal. 22 (4), 671–677.

Haxton, T.M., Uber, J., 2010. Flushing under source uncertainties. American Society of Civil Engineers (ASCE), 12th Annual Water Distribution Systems Analysis (WDSA) Conference, Tucson, AZ.

Hosni, A.A., Shane, W.T., Szabo, J.G., Bishop, P.L., 2009. The disinfection efficacy of chlorine and chlorine dioxide as disinfectants of *Bacillus globigii*, a surrogate for *Bacillus anthracis*, in water networks: A comparative study. Can. J. Civil Eng. 36 (4), 732–737.

Hosni, A.A., Szabo, J.G., Bishop, P.L., 2011. Efficacy of chlorine dioxide as a disinfectant for *Bacillus* spores in drinking-water biofilms. J Environ. Eng. ASCE 137 (7), 569–574.

Kauppinen, A., Ikonen, J., Pursiainen, A., Pitkanen, T., Miettinen, I.T., 2012. Decontamination of a drinking water pipeline system contaminated with adenovirus and *Escherichia coli* utilizing peracetic acid and chlorine. J Water Health 10 (3), 406–418.

Kirmeyer, G.J., 2000. Guidance manual for maintaining distribution system water quality. American Water Works Association (AWWA), Denver, CO.

Kirmeyer, G.J., Friedman, M., Martel, K., Howie, D., LeChevallier, M., Abbaszadegan, M., Karim, M., Funk, J., Harbour, J., 2001. Pathogen intrusion into the distribution system. American Water Works Association, Denver, CO.

Kitney, P., Woulfe, R., Codd, S., 2001. Air scouring of water mains: An asset management approach. 64th Annual Water Industry Engineers and Operators' Conference, Bendigo, Australia, 48–56.

Lapos, B.M., Brachman, R.W.I., Moore, I.D., 2007. Response to overburden pressure of an HDPE pipe pulled in place by pipe bursting. Can. Geotechnical J. 44, 957–965.

Love, A.H., Bailey, C.G., Hanna, M.L., Hok, S., Vu, A.K., Reutter, D.J., Raber, E., 2011. Efficacy of liquid and foam decontamination technologies for chemical warfare agents on indoor surfaces. J. Hazard. Mater. 196, 115–122.

Mains, C., 2008. Biofilm control in distribution systems. Tech. Brief. 8 (2), 1–4. National Environmental Services Center, Morgantown, WV.

Morrow, J.B., Almeida, J.L., Fitzgerald, L.A., Cole, K.D., 2008. Association and decontamination of *Bacillus* spores in a simulated drinking water system. Water Res. 42 (20), 5011–5021.

Najafi, M., Gokhale, S.B., 2005. Trenchless technology: Pipeline and utility design, construction, and renewa. McGraw-Hill Companies, New York.

National Institute for Occupational Safety and Health (NIOSH), 1985. Occupational safety and health guidance manual for hazardous waste site activities. Prepared by: National Institute for Occupational Safety and Health (NIOSH), Occupational Safety and Health Administration (OSHA), U.S. Coast Guard (USCG), U.S. Environmental Protection Agency (EPA). NIOSH, Washington, DC.

O'Reilly, M., Stovin, V., 1996. Trenchless construction: Risk assessment and management. Tunnelling Underground Space Technol. 11, 25–35.

Packard, B.H., Kupferle, M.J., 2010. Evaluation of surface sampling techniques for collection of *Bacillus* spores on common drinking water pipe materials. J. Environ. Monit. 12 (11), 361–368.

Porco, J.W., 2010. Municipal water distribution system security study: Recommendations for science and technology investments. J. Am. Water Works Assoc. 102 (4), 30–32.

Poulin, A., Mailhot, A., Grondin, P., Delorme, L., Periche, N., Villeneuve, J.P., 2008. Heuristic approach for operational response to drinking water contamination. J. Water Res. Plann. Manage. ASCE 134 (5), 457–465.

Poulin, A., Mailhot, A., Periche, N., Delorme, L., Villeneuve, J.P., 2010. Planning unidirectional flushing operations as a response to drinking water distribution system contamination. J. Water Res. Plann. Manage. ASCE 136 (6), 647–657.

Quarini, G., Ainslie, E., Herbert, M., Deans, T., Ash, D., Rhys, D., Haskins, N., Norton, G., Andrews, S., Smith, M., 2010. Investigation and development of an innovative pigging technique for the water-supply industry. Proceedings of the Institution of Mechanical Engineers Part E. J. Process Mech. Eng. 224 (E2), 79–89.

Raber, E., Burklund, A., 2010. Decontamination options for *Bacillus anthracis*-contaminated drinking water determined from spore surrogate studies. Appl. Environ. Microbiol. 76 (19), 6631–6638.

Radosavljevic, V., Belojevic, G., 2009. A new model of bioterrorism risk assessment. Biosecurity Bioterrorism-Biodefense Strategy Pract. Sci. 7 (4), 443–451.

Rose, L.J., O'Connell, H., 2009. UV light inactivation of bacterial biothreat agents. Appl. Environ. Microbiol. 75 (9), 2987–2990.

Satterfield, Z., 2007. Line pigging. Tech. Brief. 7 (1). National Environmental Services Center, Morgantown, WV.

Selvakumar, A., Clark, R.M., Sivaganesan, M., 2002. Costs for water supply distribution system rehabilitation. J. Water Res. Plann. Manage. ASCE 128 (4), 303–306.

Stewart, P.S., Rayner, J., Roe, F., Rees, W.M., 2001. Biofilm penetration and disinfection efficacy of alkaline hypochlorite and chlorosulfamates. J. Appl. Microbiol. 91 (3), 525–532.

Szabo, J.G., Impellitteri, C.A., Govindaswamy, S., Hall, J.S., 2009. Persistence and decontamination of surrogate radioisotopes in a model drinking water distribution system. Water Res. 43 (20), 5004–5014.

Szabo, J.G., Muhammad, N., Heckman, L., Rice, E.W., Hall, J., 2012. Germinant-enhanced decontamination of *Bacillus* spores adhered to iron and cement-mortar drinking water infrastructures. Appl. Environ. Microbiol. 78 (7), 2449–2451.

Szabo, J.G., Rice, E.W., Bishop, P.L., 2006. Persistence of *Klebsiella pneumoniae* on simulated biofilm in a model drinking water system. Environ. Sci. Technol. 40 (16), 4996–5002.

Szabo, J.G., Rice, E.W., Bishop, P.L., 2007. Persistence and decontamination of *Bacillus atrophaeus* subsp *globigii* spores on corroded iron in a model drinking water system. Appl. Environ. Microbiol. 73 (8), 2451–2457.

Tighe, S., Knight, M., Papoutsis, D., Rodriguez, V., Walker, C., 2002. User cost savings in eliminating pavement excavations through employing trenchless technologies. Can. J. Civil Eng. 29 (5), 751–761.

U.S. Environmental Protection Agency (U.S. EPA), 2001. The incorporation of water treatment effects on pesticide removal and transformations in Food Quality Protection Act (FQPA) drinking water assessments. EPA, Office of Pesticide Programs, Washington, DC.

U.S. Environmental Protection Agency (U.S. EPA), 2002a. Finished water storage facilities, Office of Water (4601M). EPA, Washington, DC.

U.S. Environmental Protection Agency (U.S. EPA), 2002b. Guidance for water utility response, recovery, and remediation actions for man-made and/or technological emergencies, EPA 810-R-02-001. EPA, Washington, DC.

U.S. Environmental Protection Agency (U.S. EPA), 2002c. Guidance for quality assurance project plans, EPA QA/G-5. EPA, Washington, DC.

U.S. Environmental Protection Agency (U.S. EPA), 2003a. Module 6: remediation and recovery guide, EPA-817-D-03–006. Response Protocol Toolbox (RPTB) interim final: Planning for and responding to contamination threats to drinking water systems. EPA, Washington, DC.

U.S. Environmental Protection Agency (U.S. EPA), 2003b. Planning for and responding to drinking water contamination treats and incidents: overview and application, EPA-817-D-03–007. Response Protocol Toolbox (RPTB) interim final: Planning for and responding to contamination threats to drinking water systems. EPA, Washington, DC.

U.S. Environmental Protection Agency (U.S. EPA), 2003c. Using the triad approach to streamline Brownfields site assessment and cleanup. Brownfields technology primer series. EPA, Washington, DC.

U.S. Environmental Protection Agency (U.S. EPA), 2007. Water contaminant information tool, EPA 817-F-07-001. EPA, Washington, DC.

U.S. Environmental Protection Agency (U.S. EPA), 2011. Report on the 2010 U.S. Environmental Protection Agency (EPA) Decontamination Research and Development Conference, EPA/600/R-11/052. National Homeland Security Research Center, Cincinnati, OH.

U.S. Environmental Protection Agency (U.S. EPA), 2012a. Containment and disposal of large amounts of contaminated water: A support guide for water utilities, EPA 817-B-12-002. EPA, Washington, DC.

U.S. Environmental Protection Agency (U.S. EPA), 2012b. Laboratory resources for the water sector to support decontamination activities, EPA 817-F-12-003. EPA, Washington, DC.

U.S. Environmental Protection Agency (U.S. EPA), 2012c. Removing biological and chemical contamination from a building's plumbing system: Method development and testing, EPA/600/R-12/032. EPA, Washington, DC.

Vitanage, D., Pamminger, F., Vourtsanis, T., 2004. Maintenance and survey of distribution systems. In: Ainsworth, R. (Ed.), Safe Piped Water: Managing Microbial Water Quality in Piped Distribution Systems. IWA Publishing, London, UK.

Volchek, K., Fingas, M., Hornof, M., Boudreau, L., Yanofsky, N., 2006. Decontamination in the event of a chemical or radiological terrorist attack. In: Frolov, K.V., Baecher, G.B. (Ed.). Protection of Civilian Infrastructure from Acts of Terrorism Springer, Dordrecht, The Netherlands.

Water Sector Decontamination Working Group (WSDWG), 2008. Recommendations and proposed strategic plan: Water sector decontamination priorities. Critical Infrastructure Partnership Advisory Council (CIPAC). Washington, DC.

World Health Organization (WHO), 2002. Environmental health in emergencies and disasters: A practical guide. WHO, Geneva, Switzerland.

Zhao, J.Q., Daigle, L., 2001. Structural performance of sliplined watermain. Can. J. Civil Eng. 28 (6), 969–978.

Chapter 8

Conclusions

This book is an accumulation of ideas and approaches for advancing security in drinking water systems. Over history, few significant threats have been made against drinking water systems, although a few isolated events and crisis situation have occurred related to hurricanes, outbreaks, and other disasters. Although there are few regulatory requirements regarding water security infrastructure in the United States, the abundance of guidance documents can be overwhelming to navigate. The effort expended by the United States reflects the concern regarding water security, but also the idea that implementing water security programs is a site-specific exercise best left to the individual water utilities. Multiple programs have been established to assist water utilities in ensuring secure drinking water systems.

A few themes stand out in this book. One is that emergency planning is essential for water security. Water utilities must have a good understanding of system components and vulnerabilities. Clear protocols must be established for evaluating and responding to water system threats. Developing good communication with partnering agencies and groups is beneficial and should be initiated during planning. Next, training and exercises are a necessary component of a comprehensive emergency response program, so that water utility employees and other responders are prepared for threats. Many devices can be used to physically protect water system equipment and facilities. Careful planning and implementation of these security measures should be undertaken, using a cost-conscience approach. Incorporation of security features can be accomplished by building these into new facilities and improvement projects.

A shift in water system management that needs to occur is to have better monitoring of water quality in distribution systems, which is an important part of a contamination warning system (CWS). Although advanced sensors are available and still emerging, monitoring a suite of water quality parameters has been shown to be protective of many contamination events. Adoption of a sensor network approach with sensors located throughout the distribution system is an effective way for monitoring a distribution system. Water utilities have many tools to implement a CWS, including sensor technologies and software. Use of these technologies makes real-time detection and containment more feasible.

Water security improvements must offer other, dual-use benefits for systems to be sustainable. These improvements are typically costly and, if effective, show no measureable benefit. Hence, other benefits need to be realized. An example of a dual-use benefit is better security against vandals and theft at facilities. Implementation of a CWS can offer the benefits of better water quality (taste and odors) as well as better disinfectant residual and lower disinfection by-products. Use of a CWS can also assist water utilities in keeping track of operational conditions, such as locating system leaks.

In the event that a water contamination event does occur, water utilities need to be prepared. Systems to provide alternative drinking water supplies need to be put into place quickly, necessitating that these systems be planned in advance. Methods for providing water for sanitation and firefighting also need to be carried out. Rehabilitation methods are available for confirmed contamination. The organizational structure for rehabilitation and recovery of a contaminated water system is similar to that used for hazardous waste sites. Response and rehabilitation efforts are greatly improved by effective planning.

One key to water security is advancement of technology. Technology can be used to aid in all aspects of preparing for, responding to, and recovering from contamination. Specific areas for technology development identified in this book are:

- Physical prevention devices.
- Sensors that monitor water quality and indicate the presence of contaminants.
- Sensor network software used to analyze data and detect contamination.
- Rapid test kits for basic screening of water samples.
- Mobile treatment units to provide potable water, provide water for sanitation and firefighting, and treat contaminated water prior to surface water disposal.
- Non-potable water firefighting methods.
- Personal protective equipment for emergency responders.
- Robots for emergency response.
- Rehabilitation methods for restoring water system components.

By advancing technology and applying it using a defense-in-depth approach, water security nationwide can be enhanced.

List of Acronyms Used in Water Security Literature and Legislation

AA	atomic adsorption
AES	atomic emission spectrometry
AMWA	Association of Metropolitan Water Agencies
ANSI	American National Standards Institute
AWWA	American Water Works Association
AWWARF	AWWA Research Foundation
BCP	business continuity plan
BT	biological terrorism
BWC	Biological Weapons Convention
BWSN	Battle of the Water Sensor Networks
C/B	chemical/biological
CCR	consumer confidence report
CCL	Contaminant Candidate List
CCTV	closed circuit television
CD	compact disk
CDC	U.S. Centers for Disease Control and Prevention
CERCLA	Comprehensive Environmental Response, Compensation, and Liability Act
CETL	Compendium of Environmental Testing Laboratories
CFR	Code of Federal Regulations
Cfs	cubic feet per second
CHRIS	Chemical Hazards Response Information System
CIAO	Critical Infrastructure Assurance Officer
CIKR	Critical Infrastructure and Key Resources
CIMS	crisis information management software
CIPAC	Critical Infrastructure Partnership Advisory Council
CIPAG	Critical Infrastructure Protection Advisory Group
CS2SAT	Control Systems Cyber Security Self-Assessment Tool
CSM	conceptual site model
CWC	Chemical Weapons Convention

CW	chemical warfare
CWA	Clean Water Act
CWS	contamination warning system
DBP	disinfection by product
DHS	U.S. Department of Homeland Security
DNA	deoxyribonucleic acid
DO	dissolved oxygen
DOC	Department of Commerce
DOD	Department of Defense
DOI	Department of the Interior
DOJ	Department of Justice
DOL	Department of Labor
DOS	Department of State
DOT	Department of Transportation
DSRC	Distribution System Research Consortium
DSS	distribution system simulator
DWSRF	Drinking Water State Revolving Fund
DDBP	disinfectants and disinfections by-products
EDS	event detection system
EMAC	Emergency Management Assistance Compact
EMPACT	Environmental Monitoring for Public Access and Community Tracking
EMS	emergency medical services
ESF	emergency support function
EPS	extended period simulation
ERLN	Environmental Response Laboratory Network
ERP	emergency response plan
ESF	emergency support function
ETV	Environmental Technology Verification Program
EWS	early warning system
FBI	Federal Bureau of Investigation
FCO	federal coordinating officer
FEMA	Federal Emergency Management Agency
FERN	Food Emergency Response Network
FOIA	Freedom of Information Act
FORMS	field operations and records management system
FRMAC	Federal Radiological Management Center
fps	foot per second
FRP	Federal Response Plan
FSP	field sampling plan
FY	Fiscal year
GC	gas chromatography
GCC	Government Coordinating Council
GIS	geographical information systems

Gpm	gallons per minute
GSA	U.S. Government Services Agency
HASP	health and safety plan
HazMat	hazardous materials
HHS	Department of Health and Human Services
HPLC	high-performance liquid chromatography
HSEEP	Homeland Security Exercise and Evaluation Program
HSPD	Homeland Security Presidential Directive
IC	incident commander
ICS	incident command system
ICP	inductively coupled plasma
ID-50	median infectious dose
IDSE	initial distribution system evaluation
INL	Idaho National Laboratory
IPD	impact probability distribution
IO	information officer
ISAC	Information Sharing and Analysis Center
ISE	ion selective electrode
ISO	International Organization for Standardization
IT	information technology
JIC	Joint Information Center
JOC	Joint Operations Center
LD-50	median lethal dose
LC	liquid chromatography
LFA	lead federal agency
LO	liaison officer
LPoC	laboratory point of contact
LRN	Laboratory Response Network
MCL	maximum contaminant level
MIP	mixed integer programming
MWCO	molecular weight cut-off
NA-ACO	non-dominated archiving ant colony optimization
NACWA	National Association of Clean Water Agencies
NCP	National Oil and Hazardous Substances Pollution Contingency Plan
NDWAC	National Drinking Water Advisory Council
NEMI-CBR	National Environmental Methods Index for Chemical, Biological, and Radiological Methods
NFPA	National Fire Protection Association
NIIMS	National Interagency Incident Management System
NIMS	National Incident Management System
NIPC	National Infrastructure Protection Center
NIPP	National Infrastructure Protection Plan
NIST	National Institute of Standards and Technology

NOAEL	no observed adverse effect level
NRC	National Research Council
NRC	Nuclear Regulatory Commission
NRF	National Response Framework
NRP	National Response Plan
NRWA	National Rural Water Association
ORD	Office of Research and Development
ORP	oxygen reduction potential
OSC	on-scene coordinator
OSHA	Occupational Safety and Health Administration
OSWER	Office of Solid Waste and Emergency Response
OW	Office of Water
PAC	project advisory committee
PAO	public affairs officer
PCCIP	President's Commission on Critical Infrastructure Protection
PCR	polymerase chain reaction
PDD	Presidential Decision Directive
PE	Professional Engineer
PIN	personal identification number
PLC	programmable logic controller
PN	public notification
POTW	publicly owned treatment works
PPE	personal protective equipment
PRG	preliminary remediation goal
PSI	pounds per square inch
PWS	public water system
QA	quality assurance
QAPP	quality assurance project plan
QC	quality control
RBES	rule-based expert system
RAIS	Risk Assessment Information System
RAMCAP	Risk Analysis and Management for Critical Asset Protection
RAM-W	Risk Assessment Methodology – Water
RCRA	Resource Conservation and Recovery Act
RF	response factor
RFP	request for proposal
RG	response guideline
RLRP	regional laboratory response plan
RMSE	root mean square error
ROC	receiver operating characteristics
RPTB	Response Protocol Toolbox
RST	regional support team
RTU	remote terminal unit
SAP	sampling and analysis plan

SCADA	supervisory control and data acquisition
SDWA	Safe Drinking Water Act, as amended
SDWIS	Safe Drinking Water Information System
SEMS	Security and Environmental Management System
SLOTS	sensors local optimal transformation system
SM	Standard Methods for the Analysis of Water and Wastewater
SOP	standard operating procedure
SSP	Sector-Specific Plan
SVOC	semi-volatile organic chemical
TBD	to be determined
TDB	treatability database (U.S. EPA)
TEVA-SPOT	Threat Ensemble Vulnerability Assessment Sensor Placement Optimization Tool
TIC	total inorganic carbon
TOC	total organic carbon
TTEP	Technology Testing and Evaluation Panel
UCMR	Unregulated Contaminant Monitoring Rule
UICP	Utility Integrated Contingency Plan
UPS	uninterrupted power supply
URL	Uniform Resource Locator
USACE	United States Army Corps of Engineers
USAMRIID	United States Army Medical Research Institute of Infectious Diseases
USCG	United States Coast Guard
USDA	United States Department of Agriculture
U.S. EPA	U.S. Environmental Protection Agency
USGS	U.S. Geological Survey
UV	ultraviolet
VA	vulnerability assessment
VOC	volatile organic chemical
VSAT	Vulnerability Self-Assessment Tool
WARN	Water/Wastewater Agency Response Network
WCIT	Water Contaminant Information Tool
WaterISAC	Water Information Sharing and Analysis Center
WaterSC	Water Security Channel
WEF	Water Environment Federation
WHO	World Health Organization
WLN	Water Laboratory Network
WMD	weapons of mass destruction
WSCC	Water Sector Coordinating Council
WSI	Water Security Initiative
WSWG	Water Security Working Group
WUERM	water utility emergency response manager
WUOCM	water utility emergency operations center manager

Water Contamination Events Reported in English Language Newspapers

News reports of water system incursions were collected for the time period from December 2005 to January 2011. Approximately 80 reports were located and the results are summarized in Table B.1.

As shown in Figure B.1, most of the events occurred in the United States. However, this could be attributed to more coverage of U.S. events by the media sources included in the search.

Of the events reported in the media, the contaminant was not determined in most cases. When the contaminant was identified, miscellaneous chemicals, rat poison, and herbicides were the most common (Figure B.2).

For the surveyed water system incursions, the supply was typically affected; although it was not always clear what portion of the supply was attacked (Figure B.3). Reservoirs and storage tanks proved to be common targets.

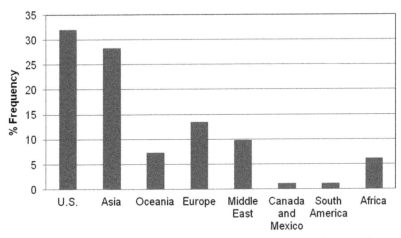

FIGURE B.1 Locations of water system incursions reported in news media (12/05 to 1/11).

TABLE B.1 Water Contamination Events as Described in the News Media from December 2005 to January 2011

Date	Location	Description of Event	Cause/Motivation
Dec 05	Fairview Township, PA	Fluoride spill at water treatment plant	Disgruntled employee
Dec 05	Azerbaijan	Suspected intent to contaminate reservoir	Unknown motives
Mar 06	Blackstone, MA	Contamination of water storage facility	Vandalism
Jun 06	Rock Creek, OH	Water reservoirs contaminated	Vandalism
Jun 06	Bowen, Australia	Contamination of water tank with herbicide (glyphosate)	Unknown
Aug 06	Tring, United Kingdom	Contamination of reservoir with weed killer (sodium chlorate)	Unknown
Oct 06	Greve, Denmark	Contamination of well with rat poison (strychnine)	Unknown
Oct 06	San Luis Obispo, CA	Contamination of private well with human waste	Neighbor dispute
Oct 06	Harrisburg, PA	Hackers broke into water treatment plant computer system	Unknown
Nov 06	Goose Creek, SC	Teacher's coffee contaminated with methanol and antifreeze (ethylene glycol)	Unknown
Jan 07	Endicott, NY	Contamination of wells with antifreeze (ethylene glycol) from skating rink	Unknown
Apr 07	Spencer, MA	Water treatment plant released sodium hydroxide into water supply	Accident
Apr 07	Harbin, China	Mass food poisoning at hospital, rice cooked with contaminated water, rat poison (fluoroacetamide)	Unknown
Apr 07	Camberwell, Australia	Salmonellosis outbreak at nursing home	Unknown

Sep 07	Letcher Co., KY	Contamination of reservoir	Vandalism
Sep 07	Sana, Yemen	Plan to poison water reservoirs at police and military posts	Support Shiite rebels
Sep 07	Wooler, United Kingdom	Terrorist threat to poison water	Unknown
Apr 08	Woodland Park, CO	Bodies of three ground squirrels found in 80,000 gallon drinking water storage tank	Unknown
May 08	Las Lomas, CA	High levels of mercury found in water supply tank	Vandalism
Jul 08	Northants, United Kingdom	Rabbit infected with parasites found in water tank	Possibly deliberate
Jul 08	Guilin, Guangxi, China	School's water storage tank smelled of pesticide and had an ivory-white substance	Possibly deliberate
Aug 08	Casterton, Victoria, Australia	Dead sheep in a carving club's water supply	Possibly deliberate
Aug 08	Unnamed major city	Plan to poison the water supply of unnamed major cities	Support of Islamic terrorist
Sep 08	Sri Lanka	Reports received that wells of armed forces and police could be poisoned	Unknown
Sep 08	Great Britain and Demark	Plan to poison water supply of Denmark, Great Britain, and other European countries	Retaliate to Denmark's mockery of Prophet Muhammad
Sep 08	New York, NY	Teacher poisoned by drinking from water bottle spiked by student with calcium hydroxide	Unknown
Sep 08	Orissa, India	Attempt to poison water at relief camps	Religious terrorists
Oct 08	Riyadh, Saudi Arabia	Plan to use cyanide to poison water supply	Al-Qaeda terrorist attack

(Continued)

TABLE B.1 Water Contamination Events as Described in the News Media from December 2005 to January 2011—Cont'd

Date	Location	Description	Cause
Oct 08	Nicosia, Cyprus	Poachers poisoned hundreds of captivity-bred partridges in retaliation for recent poaching crackdowns	Retaliation
Oct 08	Varney, WV	Plan to contaminate water supply with cyanide	Unknown
Nov 08	San Francisco, CA	Colleague's drinking water poisoned with a chemical buffer agent	Unknown
Nov 08	Iraq	Plan to poison water supply with nitric acid	Al-Qaeda terrorist attack
Dec 08	Swaziland	Contamination of water tank with poisonous substance	Property dispute
Jan 09	Auckland, New Zealand	Warning received that a water reservoir was contaminated with cyanide	Unknown
Mar 09	Moscow, Russia	Plan to poison city water supply	Terrorist attack
Apr 09	Mae Sot, Thailand	Herbicide used to poison water supply at refugee camp	Unknown
Apr 09	Santa Clarita, CA	"Bleachlike substance" found in water bottles obtained from school vending machines	Unknown
Apr 09	Western nations, in particular USA	Plan to poison water supply	Terrorist attack
Apr 09	Gandhinagar, India	Attempt to poison senior bureaucrats by loading a water cooler with insecticides	Unknown
Apr 09	Sahab, Jordan	American Al-Qaeda operative states that western economies are on the brink of failure	Unknown
Apr 09	Gujarat, India	Water cooler laced with deadly antitermite pesticide in attempt to kill employees of the Gujarat state secretariat	Unknown

Date	Location	Description	Motive
May 09	Johannesburg, South Africa	Arrested for suspicion of water contamination, the men said they dropped Orgonite pieces to improve the "etheric energy" of the dam	Believed they were helping
May 09	Taichung City, Taiwan	A man threatened universities that he would contaminate their water supply with a chlorine bomb, a bag of rat poison later found	Extortion plot
Jun 09	Maguindanao, Philippines	Rebels poisoned soldiers' sources of water	Retaliation for having camp taken
Jun 09	Geelong, Victoria, Australia	Bottle left outside door with threat, test results unknown	Unknown
Jul 09	Bauchi, Nigeria	Zealots poison water wells with phosphorous and cyanide	Political dispute
Aug 09	Maple Ridge, British Columbia	Break - in at watershed, did not seem that anything was put in water	Unknown
Aug,09	Tbilisi, Georgia	Dispute over history of previous battle between Georgia and Russia	Unknown
Sep,09	Yavatmal, India	Water tank supplying water to 4 villages poisoned	Possible revenge on rivals
Oct,09	Owenton, KY	Bombing of water pipeline	Unknown/domestic terrorism
Nov,09	New Delhi, India	Water bodies poisoned to cause harm to the paramilitary forces	Unknown
Nov,09	Minas Gerais, Brazil	Rat poison in school water supply	Unknown
Nov,09	Kaiga, India	Radioactive tritium placed in water cooler at nuclear power plant	Unknown
Nov,09	Waltham, MA	3 students place cleaning solution in a water bottle, then gave the bottle to the teacher to drink	Unknown

(Continued)

TABLE B.1 Water Contamination Events as Described in the News Media from December 2005 to January 2011—Cont'd

Nov,09	Islamabad, Pakistan	Pakistani Taliban threaten to contaminate water sources or reservoirs	Pressure army to stop military operations against them
Nov,09	Multan, Pakistan	Terrorist threat to poison water	Terrorist attack
Dec,09	Yunnan, China	Rat poison in school water	Revenge on students who tricked him
Unknown	Lupane, Zimbabwe	Borehole dismantled to deny neighboring villagers water	Property dispute
Jan 10	Cambridge, MA	Sodium azide found in the water tank of a coffee machine at Harvard Medical Schools research building	Unknown
Feb 10	Milan, Italy	Oil and diesel spill in Italy's largest river	Possible disgruntled employee
Mar 10	Walterboro, SC	9-year-old student put prescription medication and possible poisons in teacher's drink	Angered by a reassignment
Apr 10	Ross, New Zealand	1080 poisoned baits found in water supply	Unknown
May 10	Burnopfield, England	Discussed poisoning water supplies used by Muslims	Terrorist attack
May 10	Kathmandu, Nepal	Women accused of selling bottled water laced with pesticides	Worker at shop
Jun 10	Osh, Kyrgyzstan	Rumors of poison added to water supply. Clash over water rights in Fergana Valley between Uzbekistan and Kyrgyzstan	Riot
Jun 10	Bethlehem, Palestine	Man mixes rat poison into 20 liter water container used for drinking and cooking in an attempt to kill family	Unknown

Date	Location	Description	Cause
Jun 10	Toronto, Canada	Letter bombs and water bottles poisoned with hypodermic needles sent to people whom he believed "wronged" him	Revenge
Jun 10	Jind, Harayana, India	Two students poison water tank at school	Attempt to extend school break
Jul 10	Thurmont, MD	Police chief drinks bottled water and finds a 1.5 inch blue and yellow capsule in bottle. Bottle was previously sealed	Unknown
Aug 10	Tabor City, NC	Resident swimming pool contaminated with lye or drain opener	Unknown
Sep 10	Buncrana, Ireland	Poachers threaten to poison Crana River and Pollan Dam	Poachers want to continue poaching
Sep 10	Shelton, CT	Muriatic acid (a hydrochloric acid solution) poured into neighbors water system	Neighbor dispute
Sep 10	Janakpurdham, Nepal	Rumor of poisoned water in city	Unknown
Sep 10	Kalamazoo, MI	Fence at water storage facilities cut but no sign of contamination	Unknown
Oct 10	Mashonaland West, Zimbabwe	5 liters of Paraquat poured in river, fish collected, later found that 30 cattle have died from consuming water	Unknown
Oct 10	Enemy states of al-Qaeda	Possible use of rat poison to contaminate pipelines	Terrorist attack
Oct 10	Bowen, Australia	Water supply contaminated, leading to millions of dollars of crop loss. Poison used was herbicide placed in pipe system	Unknown
Nov 10	Bokaro, India	Maoist cadres poison pond close to Central Reserve Police Force camp	Unknown
Jan 11	Safed, Israel	Driver of diesel fuel tuck mistakenly attached tank hose to water intake nozzle	Negligence

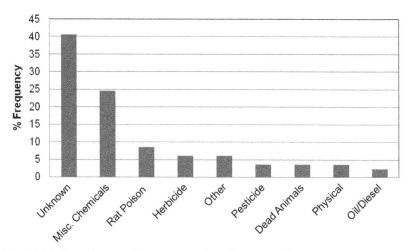

FIGURE B.2 Contaminants used in water system incursions reported in news media (12/05 to 1/11).

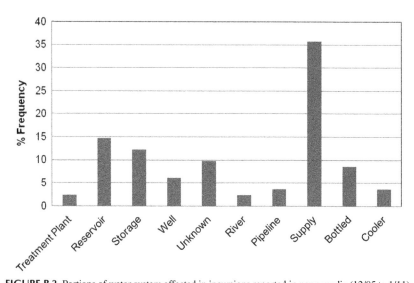

FIGURE B.3 Portions of water system affected in incursions reported in news media (12/05 to 1/11).

Chemical and Biological Threats on the Contaminant Candidate List

TABLE C.1   Data for Chemicals on the U.S. EPA Contaminant Candidate List[5]

Constituent and Use	LD-50[1] (mg/kg)	Acute Toxicity Category[2]	Water Solubility (mg/l)	Concentration Necessary for LD-50 (mg/l)[3]	Percent of Water Solubility Necessary to Reach LD-50 (%)[4]
1,1,1,2-Tetrachloroethane: Industrial chemical used in production of other substances	670	4	1,070	40,200	3,757
1,1-Dichloroethane: Industrial chemical used as a solvent	1,120	4	5,100	67,200	1,318
1,2,3-Trichloropropane: Industrial chemical used in paint manufacture	505	4	1,750	30,300	1,731
1,3-Butadiene: Industrial chemical used in rubber production	5,480	Unlikely acute toxicity	735	328,800	44,735
1,3-Dinitrobenzene: Industrial chemical used in the production of other substances	59	3	533	3,540	664
1,4-Dioxane: Solvent or solvent stabilizer used in the manufacture and processing of paper, cotton, textile products, automotive coolant, cosmetics, and shampoos	4,200	5	Miscible	252,000	—

17alpha-estradiol: Estrogenic hormone used in pharmaceuticals	980	4	0	58,800	—
1-Butanol: Used in production of other substances, as a paint solvent and food additive	790	4	73,000	47,400	65
2-Methoxyethanol: Used in consumer products: synthetic cosmetics, perfumes, fragrances, hair preparations, and skin lotions	2,460	5	Miscible	147,600	—
2-Propen-1-ol: Used in the production of other substances and to manufacture flavorings and perfumes	70	3	Miscible	4,200	—
3-Hydroxycarbofuran: Carbamate and pesticide degradate. The parent, carbofuran, is an insecticide	18	2	Miscible	1,080	—
4,4'-Methylenedianiline: Used in production of other substances and as corrosion inhibitor and curing agent for polyurethanes	120	3	1,000	7,200	720

(Continued)

TABLE C.1 Data for Chemicals on the U.S. EPA Contaminant Candidate List[5]—Cont'd

Constituent and Use	LD-50[1] (mg/kg)	Acute Toxicity Category[2]	Water Solubility (mg/l)	Concentration Necessary for LD-50 (mg/l)[3]	Percent of Water Solubility Necessary to Reach LD-50 (%)[4]
Acephate: Insecticide	945	4	818,000	56,700	7
Acetaldehyde: Used in production of other substances and as pesticide and food additive	661	4	Miscible	39,660	—
Acetamide: Used as solvent, solubilizer, plasticizer, and stabilizer	7,000	Unlikely acute toxicity	Miscible	420,000	—
Acetochlor: Herbicide for weed control on agricultural crops	763	4	233	45,780	19,648
Acetochlor ethanesulfonic acid (ESA): Acetanilide pesticide degradate. The parent, acetochlor, is used as an herbicide for weed control on agricultural crops	*	NA	Miscible	—	—
Acetochlor oxanilic acid (OA): Acetanilide pesticide degradate. The parent, acetochlor, is used as herbicide for weed control on agricultural crops	*	NA	Not found	—	—

Acrolein: Aquatic herbicide, rodenticide, and industrial chemical	10.3	2	212,000	618	0
Alachlor ethanesulfonic acid (ESA): Acetanilide pesticide degradate. The parent, alachlor, is used as an herbicide for weed control on agricultural crops	*	NA	Not found	—	—
Alachlor oxanilic acid (OA): Acetanilide pesticide degradate. The parent, alachlor, is used as an herbicide for weed control on agricultural crops	*	NA	Not found	—	—
alpha-Hexachlorocyclohexane: Component of benzene hexachloride (BHC) and formerly used as an insecticide	177	3	2	10,620	531,000
Aniline: Industrial chemical, solvent, used in the synthesis of explosives, rubber products, and in isocyanates	250	3	36,000	15,000	42
Bensulide: Herbicide	270	3	25	16,200	64,800
Benzyl chloride: Used in production of other substances: plastics, dyes, lubricants, gasoline, and pharmaceuticals	1,231	4	525	73,860	14,069

(Continued)

TABLE C.1 Data for Chemicals on the U.S. EPA Contaminant Candidate List[5]—Cont'd

Constituent and Use	LD-50[1] (mg/kg)	Acute Toxicity Category[2]	Water Solubility (mg/l)	Concentration Necessary for LD-50 (mg/l)[3]	Percent of Water Solubility Necessary to Reach LD-50 (%)[4]
Butylated hydroxyanisole: Food additive (antioxidant)	2,200	5	0	132,000	—
Captan: Fungicide	9,000	Unlikely acute toxicity	5.1	540,000	10,588,235
Chlorate (sodium chlorate): Used in agriculture as defoliant or desiccant, may occur in drinking water from disinfectants, such as chlorine dioxide	1,200	4	Miscible	72,000	—
Chloromethane (methyl chloride): Foaming agent and used in production of other substances	1,800	4	5,320	108,000	2,030
Clethodim: Herbicide	1,360	4	11.9	81,600	685,714
Cobalt: Naturally occurring element and formerly used as cobaltus chloride in medicines and as a germicide	6,170	Unlikely acute toxicity	0	370,200	—
Cumene hydroperoxide: Used as industrial chemical and in production of other substances	382	4	13,900	22,920	165

Cyanotoxins (3)*: Toxins naturally produced and released by cyanobacteria ("blue-green algae"). Various studies suggest three cyanotoxins for consideration: anatoxin-a, microcystin-LR, and cylindrospermopsin	*	NA	Not found	—	—
Dicrotophos: Insecticide	21	2	Miscible	1,260	—
Dimethipin: Herbicide and plant growth regulator	500	4	4,600	30,000	652
Dimethoate: Used as insecticide on field crops, (such as cotton), orchard crops, vegetable crops, in forestry, and for residential purposes	290	3	25,000	17,400	70
Disulfoton: Insecticide	2.3	1	16.3	138	847
Diuron: Herbicide	1,017	4	42	61,020	145,286
Equilenin Estrogenic hormone used in pharmaceuticals	*	NA	Not found	—	—

(Continued)

TABLE C.1 Data for Chemicals on the U.S. EPA Contaminant Candidate List[5] — Cont'd

Constituent and Use	LD-50[1] (mg/kg)	Acute Toxicity Category[2]	Water Solubility (mg/l)	Concentration Necessary for LD-50 (mg/l)[3]	Percent of Water Solubility Necessary to Reach LD-50 (%)[4]
Equilin: Estrogenic hormone used in pharmaceuticals	*	NA	Not found	—	—
Erythromycin: Pharmaceutical antibiotic	9,272	Unlikely acute toxicity	4.2	556,320	13,245,714
Estradiol (17-beta estradiol): Estrogenic hormone used in pharmaceuticals	*	NA	Not found	—	—
Estriol: Estrogenic hormone used in veterinary pharmaceuticals	*	NA	27.34	—	—
Estrone: Estrogenic hormone used in veterinary and human pharmaceuticals	*	NA	30	—	—
Ethinyl Estradiol (17-alpha ethynyl estradiol): Estrogenic hormone and used in veterinary and human pharmaceuticals	1,200	4	11.3	72,000	637,168

Ethoprop: Insecticide	33	2	750	1,980	264
Ethylene glycol: Antifreeze in textile manufacturing, cancelled pesticide	4,700	5	Miscible	282,000	—
Ethylene oxide: Fungicidal, insecticidal fumigant	72	3	Miscible	4,320	—
Ethylene thiourea: Used in production of other substances, such as for vulcanizing polychloroprene (neoprene) and polyacrylate rubbers, and as pesticide	545	4	20	32,700	163,500
Fenamiphos: Insecticide	2.7	1	329	162	49
Formaldehyde: Fungicide, may be a disinfection byproduct, can occur naturally	800	4	400,000	48,000	12
Germanium: Naturally occurring element, commonly used as germanium dioxide in phosphors, transistors and diodes, and in electroplating	*	NA	0	—	—

(Continued)

TABLE C.1 Data for Chemicals on the U.S. EPA Contaminant Candidate List[5]—Cont'd

Constituent and Use	LD-50[1] (mg/kg)	Acute Toxicity Category[2]	Water Solubility (mg/l)	Concentration Necessary for LD-50 (mg/l)[3]	Percent of Water Solubility Necessary to Reach LD-50 (%)[4]
Halon 1011(bromochloromethane): Used as fire-extinguishing fluid and to suppress explosions, as well as solvent in manufacturing of pesticides. May also occur as disinfection by-product in drinking water	5,000	5	16,700	300,000	1,796
HCFC-22: Used as refrigerant, as low-temperature solvent, and in fluorocarbon resins, especially in tetrafluoroethylene polymers	*	NA	Not found	—	—
Hexane: Used as solvent and is naturally occurring alkane	28,710	Unlikely acute toxicity	9.5	1,722,600	18,132,632
Hydrazine: Used in production of other substances, such as rocket propellants, and as oxygen and chlorine scavenging compound	60	3	Miscible	3,600	—

Mestranol: Estrogenic hormone and used in veterinary and human pharmaceutical	>10,000	Unlikely acute toxicity	1.132	600,000	53,003,534
Methamidophos Insecticide	14	2	Miscible	840	—
Methanol: Used as industrial solvent, gasoline additive and also as antifreeze	5,628	Unlikely acute toxicity	Miscible	337,680	—
Methyl bromide (bromomethane): Used as fumigant as fungicide	214	3	15,200	12,840	84
Methyl tert-butyl ether: Used as octane booster in gasoline, in manufacturing of isobutene, and as extraction solvent	4,000	5	51,000	240,000	471
Metolachlor: Herbicide for weed control on agricultural crops	2,200	5	530	132,000	24,906
Metolachlor ethanesulfonic acid (ESA): Acetanilide pesticide degradate. The parent, metolachlor, is used as herbicide for weed control on agricultural crops	*	NA	Not found	—	—

(Continued)

TABLE C.1 Data for Chemicals on the U.S. EPA Contaminant Candidate List[5]—Cont'd

Constituent and Use	LD-50[1] (mg/kg)	Acute Toxicity Category[2]	Water Solubility (mg/l)	Concentration Necessary for LD-50 (mg/l)[3]	Percent of Water Solubility Necessary to Reach LD-50 (%)[4]
Metolachlor oxanilic acid (OA): Acetanilide pesticide degradate. The parent, metolachlor, is used as herbicide for weed control on agricultural crops	*	NA	Not found	—	—
Molinate: Herbicide	369	4	970	22,140	2,282
Molybdenum: Naturally occurring element and commonly used as molybdenum trioxide as chemical reagent	*	NA	0	—	—
Nitrobenzene: Used in the production of aniline, as a solvent in manufacturing of paints, shoe polishes, floor polishes, metal polishes, explosives, dyes, pesticides, and drugs (acetaminophen)	600	4	2,090	36,000	1,722
Nitroglycerin: Used in pharmaceuticals, in production of explosives, and in rocket propellants	822	4	1,800	49,320	2,740

N-methyl-2-pyrrolidone: Solvent in chemical industry, and used for pesticide application and in food packaging materials	3,914	5	Miscible	234,840	—
N-nitrosodiethylamine (NDEA): Nitrosamine used as additive in gasoline and lubricants, as antioxidant, as stabilizer in plastics, and also may be disinfection by-product	220	3	106,000	13,200	12
N-nitrosodimethylamine (NDMA): Nitrosamine and formerly used in production of rocket fuels, used as industrial solvent and antioxidant, also may be disinfection by-product	27	2	Miscible	1,620	—
N-nitroso-di-n-propylamine (NDPA): Nitrosamine and may be disinfection by-product	480	4	Miscible	28,800	—
N-nitrosodiphenylamine: Nitrosamine chemical reagent used as rubber and polymer additive, may be disinfection by-product	*	NA	35.1	—	—
N-nitrosopyrrolidine (NPYR): Nitrosamine used as research chemical, may be disinfection by-product	900	4	Miscible	54,000	—

(Continued)

TABLE C.1 Data for Chemicals on the U.S. EPA Contaminant Candidate List[5]—Cont'd

Constituent and Use	LD-50[1] (mg/kg)	Acute Toxicity Category[2]	Water Solubility (mg/l)	Concentration Necessary for LD-50 (mg/l)[3]	Percent of Water Solubility Necessary to Reach LD-50 (%)[4]
Norethindrone (19-norethisterone): Progresteronic hormone used in pharmaceuticals	*	NA	7.04	—	—
n-Propylbenzene: Used in manufacture of methylstyrene, in textile dyeing, and as printing solvent, and is a constituent of asphalt and naptha	6,040	Unlikely acute toxicity	23.4	362,400	1,548,718
o-Toluidine: Used in production of other substances: dyes, rubber, pharmaceuticals and pesticides	670	4	16,600	40,200	242
Oxirane, methyl-: Industrial chemical used in the production of other substances	72	3	Miscible	4,320	—
Oxydemeton-methyl: Insecticide	48	2	Miscible	2,880	—
Oxyfluorfen: Herbicide	*	NA	0.0116	—	—

Perchlorate: Naturally occurring and human-made chemical. Used to manufacture fireworks, explosives, flares, and rocket propellant	*	NA	Not found	—	—
Perfluorooctane sulfonic acid (PFOS): Was used in firefighting foams and various surfactant uses, few of which are still ongoing because no alternatives are available	251	3	0.0031	15,060	485,806,452
Perfluorooctanoic acid (PFOA): PFOA used in manufacturing of fluoropolymers, substances that provide nonstick surfaces on cookware and waterproof, breathable membranes for clothing	*	NA	9,500	—	—
Permethrin: Insecticide	600	4	6	36,000	600,000
Profenofos: Insecticide and acaricide	400	4	28	24,000	85,714
Quinoline: Used in production of other substances, and as pharmaceutical (antimalarial) and flavoring agent	331	4	6,110	19,860	325

(Continued)

TABLE C.1 Data for Chemicals on the U.S. EPA Contaminant Candidate List[5]—Cont'd

Constituent and Use	LD-50[1] (mg/kg)	Acute Toxicity Category[2]	Water Solubility (mg/l)	Concentration Necessary for LD-50 (mg/l)[3]	Percent of Water Solubility Necessary to Reach LD-50 (%)[4]
RDX (Hexahydro-1,3, 5-trinitro-1,3,5-triazine): Explosive	100	3	59.7	6,000	10,050
sec-Butylbenzene: Used as solvent for coating compositions, in organic synthesis, as plasticizer, and in surfactants	2,240	5	17.6	134,400	763,636
Strontium: Naturally occurring element and used as strontium carbonate in pyrotechnics, in steel production, as catalyst, and as lead scavenger	*	NA	Not found	—	—
Tebuconazole: Fungicide	>5,000	Unlikely acute toxicity	36	300,000	833,333
Tebufenozide: Insecticide	>5,000	Unlikely acute toxicity	0.83	300,000	36,144,578
Tellurium: Naturally occurring element and commonly used as sodium tellurite in bacteriology and medicine	83	3	0	4,980	—

Terbufos: Insecticide	2	1	5.07	120	2,367
Terbufos sulfone phosphorodithioate: Pesticide degradate. The parent, terbufos, is used as insecticide	*	NA	Not found	—	—
Thiodicarb: Insecticide	66	3	35	3,960	11,314
Thiophanate-methyl: Fungicide	6,640	Unlikely acute toxicity	26.6	398,400	1,497,744
Toluene diisocyanate: Manufacturing of plastics	4,130	5	Miscible	247,800	—
Tribufos: Insecticide and as cotton defoliant	150	3	2.3	9,000	391,304
Triethylamine: Used in the production of other substances, and as stabilizer in herbicides and pesticides, consumer products, food additives, photographic chemicals, and carpet cleaners	460	4	88,600	27,600	31
Triphenyltin hydroxide (TPTH): Pesticide	46	2	1.2	2,760	230,000

(Continued)

TABLE C.1 Data for Chemicals on the U.S. EPA Contaminant Candidate List[5]—Cont'd

Constituent and Use	LD-50[1] (mg/kg)	Acute Toxicity Category[2]	Water Solubility (mg/l)	Concentration Necessary for LD-50 (mg/l)[3]	Percent of Water Solubility Necessary to Reach LD-50 (%)[4]
Urethane: Paint ingredient	1,809	4	480,000	108,540	23
Vanadium: Naturally occurring element and commonly used as vanadium pentoxide in production of other substances and as catalyst	*	NA	0	—	—
Vinclozolin: Fungicide	10,000	Unlikely acute toxicity	2.6	600,000	23,076,923
Ziram: Fungicide	320	4	65	19,200	29,538

[1]Median lethal dose for rats, oral
[2]Categories based on United Nations Globally Harmonized System of Classification and Labelling of Chemicals (GHS) Acute Toxicity scheme for oral exposure
[3]For a 60 kg person drinking 1 L of water
[4]For a 60 kg person drinking 1 L of water. Values over 100 indicate LD-50 cannot be reached, even at saturation
[5]From the U.S. EPA Contaminant Candidate List 3 (http://water.epa.gov/scitech/drinkingwater/dws/ccl/ccl3.cfm)

TABLE C.2 Data for Chemicals on the U.S. EPA Contaminant Candidate List, Screened by Solubility and Toxicity[5]

Constituent	LD-50[1] (mg/kg)	Acute Toxicity Category[2]	Water Solubility (mg/l)	Concentration Necessary for LD-50[3] (mg/l)	Percent of Water Solubility Necessary to Reach LD-50 (%)[4]	Amount Needed to Contaminate 1 MG of Water to LD-50 Concentration[3] (kg)	(m³)
1,4-Dioxane	4,200	5	Miscible	252,000	—	953,820	923
1-Butanol	790	4	73,000	47,400	64.93	179,409	221
2-Methoxyethanol	2,460	5	Miscible	147,600	—	558,666	579
2-Propen-1-ol	70	3	Miscible	4,200	—	15,897	19
3-Hydroxycarbofuran	18	2	Miscible	1,080	—	4,088	3
Acephate	945	4	81,8000	56,700	6.93	214,610	159
Acetaldehyde	661	4	Miscible	39,660	—	150,113	192
Acrolein	10.3	2	212,000	618	0.29	2,339	3
Aniline	250	3	36,000	15,000	41.67	56,775	56
Chlorate (sodium chlorate)	1,200	4	Miscible	72,000	—	272,520	109
Dicrotophos	21	2	Miscible	1,260	—	4,769	4
Dimethoate	290	3	25,000	17,400	69.60	65,859	52
Ethylene glycol	4,700	5	Miscible	282,000	—	1,067,370	958
Ethylene oxide	72	3	Miscible	4,320	—	16,351	19
Fenamiphos	2.7	1	329	162	49.24	613	1
Formaldehyde	800	4	400,000	48,000	12.00	181,680	170

(Continued)

TABLE C.2 Data for Chemicals on the U.S. EPA Contaminant Candidate List, Screened by Solubility and Toxicity[5]—Cont'd

Constituent	LD-50[1] (mg/kg)	Acute Toxicity Category[2]	Water Solubility (mg/l)	Concentration Necessary for LD-50[3] (mg/l)	Percent of Water Solubility Necessary to Reach LD-50[4] (%)	Amount Needed to Contaminate 1 MG of Water to LD-50 Concentration[3] (kg)	(m³)
Hydrazine	60	3	Miscible	3,600	—	13,626	13
Methamidophos	14	2	Miscible	840	—	3,179	3
Methyl bromide (bromomethane)	214	3	15,200	12,840	84.47	48,599	28
N-methyl-2-pyrrolidone	3,914	5	Miscible	234,840	—	888,869	866
N-nitrosodiethylamine (NDEA)	220	3	106,000	13,200	12.45	49,962	53
N-nitrosodimethylamine (NDMA)	27	2	Miscible	1,620	—	6,132	6
N-nitroso-di-n-propylamine (NDPA)	480	4	Miscible	28,800	—	109,008	119

N-nitrosopyrrolidine (NPYR)	900	4	Miscible	54,000	—	204,390	186
Oxirane, methyl-	72	3	Miscible	4,320	—	16,351	19
Oxydemeton-methyl	48	2	Miscible	2,880	—	10,901	8
Toluene diisocyanate	4130	5	Miscible	247,800	—	937,923	769
Triethylamine	460	4	88,600	27,600	31.15	104,466	144
Urethane	1,809	4	480,000	108,540	22.61	410,824	419

[1]Median lethal dose for rats, oral

[2]Categories based on United Nations Globally Harmonized System of Classification and Labelling of Chemicals (GHS) Acute Toxicity scheme for oral exposure

[3]For a 60 kg person drinking 1 L of water

[4]For a 60 kg person drinking 1 L of water. Values over 100 indicate LD-50 cannot be reached, even at saturation

[5]From the U.S. EPA Contaminant Candidate List 3 (http://water.epa.gov/scitech/drinkingwater/dws/ccl/ccl3.cfm)

TABLE C.3 Descriptions of Microbiological Contaminants on the U.S. EPA Contaminant Candidate List[1]

Microbial Contaminant Name	Information
Adenovirus	Virus; causes respiratory and occasionally gastrointestinal illness
Caliciviruses	Virus (including Norovirus); causes gastrointestinal illness
Campylobacter jejuni	Bacterium; causes gastroentestinal illness
Enterovirus	Viruses including polioviruses, coxsackieviruses, and echoviruses; causes mild respiratory illness
Escherichia coli (O157:H7)	Bacterium; Produces toxins and causes gastrointestinal illness and kidney failure
Helicobacter pylori	Bacterium; causes ulcers and cancer
Hepatitis A virus	Virus; causes liver disease and jaundice
Legionella pneumophila	Bacterium; causes lung disease and found in environments such as hot water systems
Mycobacterium avium	Bacterium; causes lung infection in immunocompromised and those with lung disease
Naegleria fowleri	Protozoan parasite; causes primary amebic meningoencephalitis and found in warm surface and ground water
Salmonella enterica	Bacterium; causes gastrointestinal illness
Shigella sonnei	Bacterium; causes gastrointestinal illness and bloody diarrhea

[1]From the U.S. EPA Contaminant Candidate List 3 (http://water.epa.gov/scitech/drinkingwater/dws/ccl/ccl3.cfm)

Physical Prevention Devices for Water Security

TABLE D.1 Devices and Equipment Used for Water System Hardening

Manufacturer and Website	Name of Product	Description
Alarms		
RACO Manufacturing, Inc. www.racoman.com	Verbatim Alarm Monitoring Control System (VSS Models)	Autodialer, remote, supervisory, SCADA, PLC network interface; monitors flow, level, pressure, temperature, pH, etc.
Sensaphone, Inc. www.sensaphone.com	Sensaphone 1104 Alarm Monitoring and Control System	Telemetry, remote sensing, security code access; monitors power, temperature, smoke, humidity, etc.
Aboveground, outdoor equipment enclosures		
HUBBELL Power Systems; Hot Box®, Glass Pad™ www.hubbellpowersystems.com	Hot Box®, Hot Rok®, Lok Box®, Lok Rok®	Drop over enclosure; heated or unheated; aluminum and fiberglass, insulated; fiberglass pad
Security barriers		
B&B ARMR Corporation www.bb-armr.com	Sliding gate	Active barrier
Secure USA, Inc. www.secureusa.net	SU-BX Bollards	Passive barrier
Backflow prevention devices		
WATTS www.watts.com	Multiple models (e.g. GoldenEagle 919)	Health hazard applications; double check valve assemblies; dual checks with atmospheric vent

Card identification, access, tracking systems

Company	Product	Description
Datawatch Systems www.datawatchsystems.com		Proximity card
HID Corporation www.hidglobal.com	SensorCard II	Wiegand card with magnetic stripe
Fargo Electronics, Inc. (by HID) www.fargo.com	Smartcard	Magnetic stripe card

Intrusion sensors

Exterior intrusion sensors

Company	Product	Description
Magal-Senstar by Senstar www.senstar.com	OmniTrax®	Buried cable intrusion detection sensor and others
Southwest Microwave, Inc. www.southwestmicrowave.com	Micronet™, Model 455 Outdoor Active Infrared Intrusion Sensor	Perimeter fence detection system, infrared sensors, buried cable detection, alarms, microwave sensors, etc.
Integrated Security Corporation www.integratedsecuritycorp.com	Infinity Taut-wire, Infinity Passive Infrared	

Fence-associated exterior intrusion sensors

Company	Product	Description
Integrated Security Corporation www.integratedsecuritycorp.com	Infinity 2000 Electromechanical Vibration Sensor	
Safeguards Technology, Inc. www.safeguards.com	STI-290B Fence Vibration System	
Fiber SenSys, Inc. www.fibersensys.com	SC Fiber Optic Cable Sensor System	

(Continued)

TABLE D.1 Devices and Equipment Used for Water System Hardening—Cont'd

Manufacturer and Website	Name of Product	Description
Magal Security Systems Ltd. www.magal-s3.com	DTR-2000, Taut Wire Intrusion Detection System	Many other options available, refer to website
Interior intrusion sensors		
Potter Electric Signal Company www.pottersignal.com	EVD-1 Electronic Vibration System	Boundary-penetration vibration sensor
Sensaphone, Inc. www.sensaphone.com	Sensaphone 1400	Passive infrared motion detector, magnetic reed switch, boundary-penetration electromechanical sensor, smoke detector, humidistat, etc.
GE Interlogix www.ge.com	RCR/REX; Series ShatterPro III Glassbreak Sensor	Acoustic glass-break sensor; fits large or small rooms; alarm memory
Manhole intrusion sensors		
Woven Electronics www.wovenelectronics.com	LightLOC™ Sensor - SmartSwitch	Fiber optic systems manhole sensor; adjustable bracketing
CGM Security Solutions	Security Sensor	Magnetic systems manhole sensor
Hardwire	LightGuard, LightLoc	LightGuard requires physical break in contact within the sensor; LightLoc requires change in orientation of the fiber
Pure Technologies www.puretechnologiesltd.com	Soundprint	Infrared systems manhole sensor

Locks and other security devices

Door security

Maximum Security Products Corporation www.maximumsecuritycorp.com	Hollow STEEL door; Model 850, Model 855	Secure door
McKinney www.mckinneyhinge.com	T4A3786MM (five-knuckle)	Door hinge
Ceco www.cecodoor.com	ArmorShield Door and Frame System	Bullet resistant door

Fire hydrant locks

Flow Security Systems www.spinsecure.us	Captivater ™	Fire hydrant lock
McGard www.mcgard.com		Hydrant locks; nozzle locks
Mueller Company www.muellercompany.com	Hydrant DefenderT	Hydrant lock

Fire hydrant security devices

Davidson Hydrant Technologies, Inc. http://davidsonhydrant.com	Davidson Check Valve	Prevents backflow and access to hydrant barrel

(Continued)

TABLE D.1 Devices and Equipment Used for Water System Hardening—Cont'd

Manufacturer and Website	Name of Product	Description
Hatch security		
Halliday Products www.hallidayproducts.com	T-316 Slam lock; Keyed cylinder lock	Flush mounted locking system
USF Fabrication, Inc. www.usffab.com	Pentahead Bolt Lock	Bolt lock
Stabiloc™ LLC www.stabiloc.com	Bolt lock plus receiver	Bolt lock
McGard www.mcgard.com	Intimidator Man-Lock	Bolt-type hatch lock
Ladder access control		
RB Industries www.laddergate.com	Ladder Gate® Climb Preventive Shield	Aluminum ladder cover; installed a minimum of 10 feet above the ground
BROCK® www.brockgrain.com	Ladder Security Door	Hinged ladder cover, all-galvanized steel, 76 × 18.75 inches
Carbis, Inc. www.carbis.net	Security Cover, Security Access Ladder	Fixed caged ladders
Sermi Products, Inc. www.sermi.com	Removable Ladder Cover	

Manhole locks		
McGard, Inc. www.mcgard.com	Intimidator Man-Lock™	Manhole locks
Stabiloc™ LLC www.stabiloc.com	Stabiloc™ Model 911	Manhole lock
Henkels & McCoy www.henkels.com	No Access™	Pan-type manhole lock
Barton Southern Company www.bartonsouthern.com www.lockdownsolutions.com	LockDown-LockDry™	Manhole lock
Valve lockout devices		
McGard www.mcgard.com	Intimidator Valve Box-Lock™	Straight-through and right angle valves, captures shutoff valve and coupling nut
Security for vents		
ARC3 Corporation www.arcthree.com	OMEGA Vent Security Shroud™	Interior baffling system prevents intentional introduction of agents

(Continued)

TABLE D.1   Devices and Equipment Used for Water System Hardening—Cont'd

Manufacturer and Website	Name of Product	Description
Fences		
Riverdale Mills Corporation www.riverdale.com www.wirewall.com	WireWall™	Welded wire; anticlimb (small openings); anticut (robust wire and welded joints); better visibility (flat, 2D profile); different mesh, framework, coating, and color options
Films for glass shatter protection		
ShatterGARD® www.shattergard.com	BurglarGARD™, BlastGARD™	Applied directly to interior side of window/dowpane
Reservoir covers		
Temcor, Inc. www.temcor.com	Aluminum Dome Reservoir Cover	Eliminates need for roof columns and extensive reinforcement of tank walls, panels can be removed
C.W. Neal Corporation www.cwneal.com	Floating Geomembrane Cover	Reinforced and unreinforced polypropylene, high-density polyethylene, Hypalon®, PVC, Enviro Liner, and reinforced Elvaloy® liners

Detection Devices for Water Security

TABLE E.1 Available Detection Systems for Monitoring Constituents Used to Improve Water Security

Sensor	Description	Manufacturer	Website	Parameters Monitored	Reference
Optical					
Biosentry™	Online laser-based multiple-angle light scattering (MALS)	JMAR Technologies, San Diego, CA		Particle counts, size, and shapes (rods, spores, cysts)	U.S. EPA, 2010; U.S. EPA, 2009; Storey, 2011
FlowCAM®	Online flow cytometer and microscope	Fluid Imaging Technologies, Yarmouth, ME	www.fluidimaging.com	Particle size distribution	U.S. EPA, 2010; U.S. EPA, 2009
s::can spectro::lyser™ UV or UV-Vis	Online UV-Vis spectrometry	s::can Meßtechnik GmbH, Vienna, Austria	www.s-can.at	NO_3-N, COD, BOD, TOC, DOC, UV_{254}, NO_2-N, BTX, AOC, temperature, pressure, TSS, turbidity, color, O_3, H_2S (depending on application)	U.S. EPA, 2010; U.S. EPA, 2009; Storey, 2011
Hach FilterTrak™ 660 sc Laser Nephelomete	Online laser turbidimeter	Hach, Loveland, CO	www.hach.com	Turbidity	U.S. EPA, 2010; U.S. EPA, 2009
Real UVT online monitor	Online ultraviolet light transmission at 254 nm wavelength	Real Tech, Inc., Whitby, ON, Canada	www.realtech.ca		U.S. EPA, 2010; U.S. EPA, 2009

Biosensers

TOXcontrol™ (microLAN)	Real-time biosensor using luminescent bacterium	TOXcontrol	www.toxcontrol.com	Toxicity, TOC, BOD, turbidity, SAC$_{254}$, NO$_3$-N	Zurita et al., 2007; Mons, 2008; Storey, 2011
Algae Toximeter (BBE)	Biosensor using green algae	bbe	www.bbe-moldaenke.de	Toxicity, algae	de Hoogh et al., 2006; Mons, 2008; Storey, 2011
Daphnia Toximeter™ (DaphTox II)	Biosensor using water fleas *Daphnia magna*	bbe	www.bbe-moldaenke.de	Toxicity	de Hoogh et al., 2006; Jeon et al. 2008; Mons, 2008; Storey, 2011
ToxProtect 64 (BBE)	Biosensor using fish	bbe	www.bbe-moldaenke.de	Toxicity	van der Gaag and Volz, 2008; Mons, 2008; Storey, 2011
Microtox® Model 500 Analyzer	Biosensor using luminescent bacteria (nonportable)	SDIX (Strategic Diagnostics, Inc.)	www.sdix.com	Toxicity	Bulich, 1979; States et al., 2003; van der Schalie, 2006
DeltaTox II	Biosensor using luminescent bacteria (portable)	SDIX (Strategic Diagnostics, Inc.)	www.sdix.com	Toxicity, ATP	

(Continued)

TABLE E.1 Available Detection Systems for Monitoring Constituents Used to Improve Water Security—Cont'd

Sensor	Description	Manufacturer	Website	Parameters Monitored	Reference
Aquaverify	Biosensor using luminescent bacteria (nonportable)	Checklight Ltd.	www.checklight.biz	Toxicity	Ulitzur et al., 2002; van der Schalie, 2006
ToxScreen (II)[3]	Biosensor using luminescent bacteria	Checklight, Ltd.	www.checklight.biz	Toxicity	Ulitzur et al., 2002; van der Schalie, 2006
U.S. Army Center for Environmental Health Research aquatic biomonitor	Biosensor using aquatic organism ventilation characteristics	U.S. Army Center for Environmental Health Research	www.flcmidatlantic.org /pdf/awards/2005/Apparatus_and_Method.pdf	Toxicity	van der Schalie, 2004
QwikLite 200 Biosensor System	Biosensor using luminescent plankton	QwikLite®/ Assure Controls	www.assurecontrols. com/products	Toxicity	
Algorithm					
Hach Event Monitor (Guardian Blue™)	Indicates a contamination event using baseline data and predetermined triggers	Hach	www.hach.com	pH, conductivity, organic carbon, temperature, turbidity, chlorine residual (free or total), pressure	Hohman, 2007; Storey, 2011

Enzyme

Eclox	Chemiluminescent oxidation-reduction reaction catalyzed by plant enzyme	Severn Trent Services	www.hach.com	Toxicity, pH, total dissolved solids, color chlorine content, mustard gas (optional)	Hayes and Smith, 1996; States et al., 2003; van der Schalie, 2006
Mitoscan	Enzyme activity of submitochondrial particles (SMP)	Harvard Bioscience, Inc., Aquatox Reseach, Inc.	www.aquatoxresearch.com	Toxicity	Knobeloch et al., 1990; van der Schalie, 2006
Toxi-ChromoTest	Bacterial inhibition of enzyme synthesis (microplate)	Environmental Biodetection Products Inc. (ebpi)	www.ebpi-kits.com/TOXI-ChromoTest.html	Toxicity	Reinhartz et al., 1987; van der Schalie, 2006
BAX®	Polymerase chain reaction (PCR) with automated fluorescent melting curve analysis	Dupont Qualicon	www2.dupont.com/Qualicon/en_US	E coli O157:H7	Bukhari, 2007

Bioelectric

CANARY Bioelectronic Sensor	Engineered cells emit photons that indicate binding with pathogens of interest	Lincoln Laboratory, MIT	www.ll.mit.edu	Microorganisms	

(Continued)

TABLE E.1 Available Detection Systems for Monitoring Constituents Used to Improve Water Security—Cont'd

Sensor	Description	Manufacturer	Website	Parameters Monitored	Reference
Biochemical					
Reveal®	Immunological assay	Neogen Corp	www.neogen.com	E coli O157:H7	Bukhari, 2007
ImmunoCard STAT!®	Immunological assay	Meridian Bioscience, Inc.	www.meridianbioscience.com	E coli O157:H7	Bukhari, 2007
Eclipse™	Immunological assay	Eichrom Technologies Inc.	www.eprogen.com	E coli O157:H7	Bukhari, 2007
E. coli O157 Antigen Detection Test	Immunological assay	Diagnostic Automation/Cortez Diagnostics, Inc.	www.rapidtest.com	E coli O157:H7	Bukhari, 2007
Electrochemical					
Censar® Six-CENSE™	Electrochemical technology on ceramic chip	AES Global, Inc.	www.aesglobal.com	Color, turbidity, free chlorine, mono-chloramine, dissolved oxygen, pH, conductivity, oxidation-reduction potential, and temperature	USEPA, 2009; Storey, 2011

Other

YSI Sonde™ (6600V2-4)	Optical sensor	YSI	www.ysi.com	Conductivity, salinity, temperature, depth, pH, dissolved oxygen, turbidity, chlorophyll, blue-green algae, oxidation-reduction potential, total dissolved solids, resistivity	Atkinson and Mabe, 2006; Storey, 2011
Water Canary	Spectral technology	Water Canary	www.watercanary.com	Microorganisms	
MicroChemLab	Fluorescently labeled proteins on microfluidic chip according to molecular weight by capillary gel electrophoresis and mass/charge ratio using capillary zone electrophoresis, proteins detected using laser-induced fluorescence. Samples concentrated using miniaturized gas chromatography column. Detection using surface acoustic wave sensors	Sandia National Laboratories	www.sandia.gov	Biological, chemical, proteins, biomolecules, toxic industrial chemicals	

(Continued)

TABLE E.1 Available Detection Systems for Monitoring Constituents Used to Improve Water Security—Cont'd

Sensor	Description	Manufacturer	Website	Parameters Monitored	Reference
Electric Cell-Substrate Impedance Sensing (ECIS) Biosensor	Portable, automated, benchtop mammalian cell-based toxicity sensor using fluidic biochips containing endothelial cells monitored by ECIS (measured toxicant-induces changes in electrical impedance of cell monolayer)	Agave Biosystems Inc., VEC Technologies, Applied BioPhysics	www.agavebio.com www.biophysics.com	Toxicity, pathogens, chemical toxins	Curtis, 2009; Giaever and Keese, 1993; van der Schalie, 2006
MEDA-5T	FS-5T scaler analyzer, PGS-3L scintillation detector	Technical Associates	http://www.tech-associates.com	Radioactive material	
OLM-100	MD55EV1 detector	Canberra	www.canberra.com	Liquid and gas radioactivity	
SMART 3 Color-imeter	Photodiode, LED	LaMotte	http://www.lamotte.com	Cyanide	
STIPTOX-adapt (W)	Turbulent-bed biore-actor	Isco (Envitech)	http://cms.esi.info/Media/documents/Envit_toxi meter_ML.pdf	Toxicity	
Arsenic Ultra Low Quick™ II	Kit, Color Chart method, Quick™ Arsenic Scan method, Compu Scan method	Industrial Test Systems, Inc. (ITS)	http://www.sensafe.com	Arsenic	

TABLE D.2 Emerging and Developing Technologies Used to Detect Contaminants

Sensor	Description	Measures	References
Surface Enhanced Raman Spectroscopy (SERS)	Identification by spectra produced at the microbial surface following reactions with antibodies	Bacteria	Sengupta et al., 2006; van der Gaag and Volz, 2008; Storey, 2011
Laser Tweezer Raman Spectroscopy (LTRS)	Laser light forms an "optical tweezer" that catches and discriminates among different types of microorganisms	Bacteria	Xie et al., 2005; van der Gaag and Volz, 2008; Storey, 2011
Surface Acoustic Wave (SAW) Devices	Acoustic wave produces mechanical wave that travels through device	Toxicity	van der Gaag and Volz, 2008; Storey, 2011
Tyrosinase Composite Biosenser	Detection of phenol after the hydrolysis of phenyl, isopropyl for enzyme activity induction	Coliforms	Serra, 2005
Potentiometric Alternating Biosensing (PAB) Transducer	Immunoassay test, transducer based on light addressable potentiometric sensor (LAPS)	Bacteria	Ercole, 2002
Xenophus laevis melanphores cell line	Disperse melansomes following chemical exposure	Toxicity	Iuga, 2009
Fish Activity Monitoring System (FAMS)	Video surveillance using closed-circuit television (CCTV)	Toxicity	van der Gaag and Volz, 2008; Mons, 2008; Storey, 2011
Microbial Fuel Cells (MFC)	Silicon-based MFC	BOD, toxicity	Davila, 2011

(Continued)

TABLE D.2 Emerging and Developing Technologies Used to Detect Contaminants—Cont'd

Sensor	Description	Measures	References
Fiber-Optic Based Cell Biosensors	Biorecognition using cells attached to fiber optic transducer	Toxicity	Eltzov, 2010
Hepatocyte low density lipoprotein (LDL) uptake	Measures fluorescein isothiocyanate labeled hepatocyte LDL-uptake in human Hep G2 cells	Toxicity	Shoji et al., 2000; van der Schalie, 2006
Neuronal microelectrode array	Measures action potential activity in neuronal network via a noninvasive extracellular recording	Toxicity	van der Schalie, 2006
Sinorhizobium meliloti toxicity test	Bacterial mediated chemical reaction with dye indicator	Toxicity	Botsford, 2002; van der Schalie, 2006
SOS cytosensor system	Measures changes in optical appearance of fish chromatophores	Toxicity	Dierkson et al., 2004; van der Schalie, 2006

Treatment Systems for Recovery and Rehabilitation Efforts

TABLE F.1   Portable and Mobile Treatment Systems

	Manufacturer	Model	Capacity	Treatment Processes
1	TiTaN® http://www.electrochlorinator.com/	KLOROGEN® RT MOBILE CHLOR™	100–250 gal/hr 100–250 gal/hr	Electrochlorination Chlorination
2	Aqua Sun International www.aqua-sun-intl.com	Responder S Responder A Outpost S Outpost A Aqua Tender AF20 Aqua Tender AF40	1 gal/min 100 gal/d 1 gal/min 720 gal/d 300 gal/hr 720 gal/hr 24,000 gal/d 40 gal/min 48,000 gal/d	Sediment prefilter (5.0 micron), carbon block polishing filter, UV disinfection Washable/reusable filter, sediment prefilter (5.0 micron), carbon block polishing filter, UV disinfection Washable/reusable filter, sediment prefilter (5.0 micron), carbon block polishing filter, UV disinfection
3	First Water www.firstwaterinc.com	FW 60-B FW 60-S FW 720-M	60 gal/hr 60 gal/hr 12 gal/min 15,000 gal/d	Sediment filter, carbon filter, membrane filter, UV disinfection
4	Noah Water Systems www.noahwater.com	The Nomad The Trekker	25 gal/min 36,000 gal/d 1 gal/min	Filtration (pre-filter, 5-μm sediment filter, carbon filter) and UV disinfection (minimum 16,000 mW/cm² dose)

5	Global Hydration www.globalhydration.com	Can Pure™	13,209 gal/d	Dual process purification system incorporating microfiltration and chlorination
6	Water One Inc www.wateroneinc.com	TS-700M TBS-700M STS-700M WS-12M WG-12M	14 gal/min 20,000 gal/d 1.75 gal/min 2520 gal/d	ViroBac™– a proprietary media, ultraviolet light, absolute-micron filtration
7	Hammonds http://www.hammondscos.com/	Portable Disinfection Unit	150-750 gal/min	Blends bleach or other water treatment chemicals into the flow of water through the fluid motor
8	Water Treatment http://watertreatment.s2ap.ca/	Rapid Aqua Deployment System	21,000 gal/d	Advanced ozone disinfection
9	Matt Chlor Inc http://www.mattchlor.com/index.php	Portable Water Treatment Chlorination Systems	Custom	Disinfection treatment using sodium hypochlorite or chlorine gas, dechlorination, ammonia/chloramination treatment
10	RODI Systems http://www.rodisystems.com/	Portable System	Custom	Filtration, ion exchange, carbon adsorption, chemical injection, ultrafiltration, reverse osmosis, and membrane bioreactor

Index

Note: Page numbers with "f" denote figures; "t" tables.

.

Printed and bound by CPI Group (UK) Ltd, Croydon, CR0 4YY

03/10/2024

01040427-0005